TESTING & EVALUATION OF CIVIL ENGINEERING MATERIALS

AUTHOR(s)

Dr. Sandeep Chaudhary
Professor,
Department of Civil Engineering
Indian Institute of Technology Indore,
Indore

Dr. Kashyap A Patel
Assistant Professor,
Department of Civil Engineering
Sardar Vallabhbhai National Institute of Technology, Surat

REVIEWER

Dr. Sahil Bansal
Assistant Professor,
Department of Civil Engineering
Indian Institute of Technology Delhi, Delhi

KHANNA BOOK PUBLISHING CO. (P) Ltd.
4C/4344, Ansari Road, Darya Ganj,
New Delhi, India

All India Council for Technical Education
Nelson Mandela Marg, Vasant Kunj,
New Delhi, India

Dear Readers,

To prevent the piracy, this book is secured with HIGH SECURITY HOLOGRAM on the front title cover. In case you don't find the hologram on the front cover title, please write us to at contact@khannabooks.com or whatsapp us at +91-99109 09320 and avail special gift voucher for yourself.

Specimen of Hologram on front Cover title:

Moreover, there is a SPECIAL DISCOUNT COUPON for you with EVERY HOLOGRAM.

How to avail this SPECIAL DISCOUNT:

Step 1: Scratch the hologram

Step 2: Under the scratch area, your "coupon code" is available

Step 3: Logon to www.khannabooks.com

Step 4: Use your "coupon code" in the shopping cart and get your copy at a special discount

Step 5: Enjoy your reading!

ISBN: 978-93-5538-431-7

Book Code: UG424EN

Testing and Evaluation of Civil Engineering Materials

by Sandeep Chaudhary, Kashyap A Patel

First Edition: 2026

Originally Published by:

All India Council for Technical Education, Ministry of Education, GOI

Exclusively Published & Distributed by:

Khanna Book Publishing Co. (P) Ltd.

4C/4344, Ansari Road, Daryaganj,
New Delhi (INDIA)

Visit us at: **www.khannabooks.com**

Write us at: **contact@khannabooks.com**

To view complete list of books, Please scan the QR Code:

KPH

Printed in Bharat.

प्रो. टी. जी. सीताराम
अध्यक्ष
Prof. T. G. Sitharam
Chairman

सत्यमेव जयते

आज़ादी का अमृत महोत्सव

अखिल भारतीय तकनीकी शिक्षा परिषद्
(भारत सरकार का एक सांविधिक निकाय)
(शिक्षा मंत्रालय, भारत सरकार)
नेल्सन मंडेला मार्ग, वसंत कुंज, नई दिल्ली–110070
दूरभाष : 011–26131498
ई–मेल : chairman@aicte-india.org

ALL INDIA COUNCIL FOR TECHNICAL EDUCATION
(A STATUTORY BODY OF THE GOVT. OF INDIA)
(Ministry of Education, Govt. of India)
Nelson Mandela Marg, Vasant Kunj, New Delhi-110070
Phone : 011-26131498
E-mail : chairman@aicte-india.org

FOREWORD

Engineers are the backbone of any modern society. They are the ones responsible for the marvels as well as the improved quality of life across the world. Engineers have driven humanity towards greater heights in a more evolved and unprecedented manner.

The All India Council for Technical Education (AICTE), have spared no efforts towards the strengthening of the technical education in the country. AICTE is always committed towards promoting quality Technical Education to make India a modern developed nation emphasizing on the overall welfare of mankind.

An array of initiatives has been taken by AICTE in last decade which have been accelerated now by the National Education Policy (NEP) 2020. The implementation of NEP under the visionary leadership of Hon'ble Prime Minister of India envisages the provision for education in regional languages to all, thereby ensuring that every graduate becomes competent enough and is in a position to contribute towards the national growth and development through innovation & entrepreneurship.

One of the spheres where AICTE had been relentlessly working since past couple of years is providing high quality original technical contents at Under Graduate & Diploma level prepared and translated by eminent educators in various Indian languages to its aspirants. For students pursuing 2nd year of their Engineering education, AICTE has identified 87 books, which shall be translated into 12 Indian languages - Hindi, Tamil, Gujarati, Odia, Bengali, Kannada, Urdu, Punjabi, Telugu, Marathi, Assamese & Malayalam. In addition to the English medium, books in different Indian Languages are going to support the students to understand the concepts in their respective mother tongue.

On behalf of AICTE, I express sincere gratitude to all distinguished authors, reviewers and translators from the renowned institutions of high repute for their admirable contribution in a record span of time.

AICTE is confident that these outcomes based original contents shall help aspirants to master the subject with comprehension and greater ease.

T. G. Sitharam

(Prof. T. G. Sitharam)

ACKNOWLEDGEMENT

The authors are grateful to the authorities of AICTE, particularly Prof. T. G. Sitharam, Chairman; Prof. M. P. Poonia, Vice-Chairman; Prof. Rajive Kumar, Member-Secretary and Dr Amit Kumar Srivastava, Director, Faculty Development Cell for their planning to publish the books on **Testing & Evaluation of Civil Engineering Materials**. We sincerely acknowledge the valuable contributions of the reviewer of the book Dr. Sahil Bansal, Assistant Professor, IIT Delhi for making the book students' friendly and giving a better shape to the overall structure of its content.

The authors will also like to acknowledge the research scholars Dr. Akshay Anil Thakare, Mr. Sanchit Gupta and Ms. Astha Sharma of IIT Indore and the PG students Shubham, Priyank, Pawan, Tanjil, Arkoprabha, Uddeshya of SVNIT Surat for their continuous help and support during the preparation of this book.

This book is an outcome of various suggestions of AICTE members, experts and authors who shared their opinion and thought to further develop the engineering education in our country. Acknowledgements are due to the contributors and different workers in this field whose published books, review articles, papers, photographs, footnotes, references and other valuable information enriched us at the time of writing the book.

Dr. Sandeep Chaudhary

Dr. Kashyap A Patel

PREFACE

The book titled "Testing & Evaluation of Civil Engineering Materials" is an outcome of the rich experience of our teaching of both theory and lab sessions on civil engineering materials and their testing procedures. The initiation of writing this book is to expose the engineering students, to the fundamentals of various civil engineering materials and enable them to evaluate different engineering materials using standard testing procedures. Keeping in mind the purpose of wide coverage as well as to provide essential supplementary information, we have included the topics recommended by AICTE, in a very systematic and orderly manner throughout the book. Efforts have been made to explain the fundamental concepts of the subject in the simplest possible way.

During the process of preparation of this book, we have considered the various standard text books and accordingly we have developed sections like critical questions, practical, etc. While preparing the different sections emphasis has also been laid on observed behaviour of the materials, their defining characteristics and their relevance to field application. The book covers both conventionally used and emerging civil engineering materials. The book also covers sample preparation, testing procedures, gathering of relevant data and interpretation of results for evaluating various engineering materials. Among the various tests, the present book is focussed on mechanical testing of civil engineering materials, with discussion on physical and chemical. Advanced evaluation techniques like non-destructive tests, durability tests, and sustainability assessment have been covered in Know More section of the book. This book will prepare engineering students to apply the knowledge of testing and evaluation to identify and apply civil engineering materials as per the application requirements. The subject matters are presented in a constructive manner so that an Engineering degree prepares students to work in different sectors of civil engineering and adapt with the recent advances in engineering materials.

We sincerely hope that the book will inspire the students to identify, test, evaluate and apply the most suitable of engineering materials. We would be thankful to all beneficial comments and suggestions which will contribute to the improvement of the future editions of the book. It gives us immense pleasure to place this book in the hands of the teachers and students. It was indeed a big pleasure to work on different aspects covering in the book.

Dr. Sandeep Chaudhary

Dr. Kashyap A Patel

Outcome Based Education

For the implementation of an outcome based education the first requirement is to develop an outcome based curriculum and incorporate an outcome based assessment in the education system. By going through outcome based assessments evaluators will be able to evaluate whether the students have achieved the outlined standard, specific and measurable outcomes. With the proper incorporation of outcome based education there will be a definite commitment to achieve a minimum standard for all learners without giving up at any level. At the end of the programme running with the aid of outcome based education, a student will be able to arrive at the following outcomes:

PO1. **Engineering knowledge**: Apply the knowledge of mathematics, science, engineering fundamentals, and an engineering specialization to the solution of complex engineering problems.

PO2. **Problem analysis**: Identify, formulate, review research literature, and analyze complex engineering problems reaching substantiated conclusions using first principles of mathematics, natural sciences, and engineering sciences.

PO3. **Design / development of solutions**: Design solutions for complex engineering problems and design system components or processes that meet the specified needs with appropriate consideration for the public health and safety, and the cultural, societal, and environmental considerations.

PO4. **Conduct investigations of complex problems**: Use research-based knowledge and research methods including design of experiments, analysis and interpretation of data, and synthesis of the information to provide valid conclusions.

PO5. **Modern tool usage**: Create, select, and apply appropriate techniques, resources, and modern engineering and IT tools including prediction and modeling to complex engineering activities with an understanding of the limitations.

PO6. **The engineer and society**: Apply reasoning informed by the contextual knowledge to assess societal, health, safety, legal and cultural issues and the consequent responsibilities relevant to the professional engineering practice.

PO7. **Environment and sustainability**: Understand the impact of the professional engineering solutions in societal and environmental contexts, and demonstrate the knowledge of, and need for sustainable development.

PO8. **Ethics**: Apply ethical principles and commit to professional ethics and responsibilities and norms of the engineering practice.

PO9. Individual and team work: Function effectively as an individual, and as a member or leader in diverse teams, and in multidisciplinary settings.

PO10. Communication: Communicate effectively on complex engineering activities with the engineering community and with society at large, such as, being able to comprehend and write effective reports and design documentation, make effective presentations, and give and receive clear instructions.

PO11. Project management and finance: Demonstrate knowledge and understanding of the engineering and management principles and apply these to one's own work, as a member and leader in a team, to manage projects and in multidisciplinary environments.

PO12. Life-long learning: Recognize the need for, and have the preparation and ability to engage in independent and life-long learning in the broadest context of technological change.

COURSE OUTCOMES

After completion of the course the students will be able to:

CO-1: Calibrate electronic sensors

CO-2: Operate a data acquisition system

CO-3: Operate various types of testing machines

CO-4: Configure a testing machine to measure tension or compression behavior

CO-5: Compute engineering values (e.g., stress or strain) from laboratory measures

CO-6: Analyze a stress versus strain curve for modulus, yield strength and other related attributes

CO-7: Identify modes of failure

CO-8: Write a technical laboratory report

Course Outcomes	Expected Mapping with Programme Outcomes *(1- Weak Correlation; 2- Medium correlation; 3- Strong Correlation)*											
	PO-1	PO-2	PO-3	PO-4	PO-5	PO-6	PO-7	PO-8	PO-9	PO-10	PO-11	PO-12
CO-1	-	-	-	1	3	-	-	1	-	-	-	1
CO-2	-	-	-	1	3	-	-	1	1	1	-	1
CO-3	2	1	-	2	2	-	-	1	1	1	-	2
CO-4	2	1	2	2	2	1	1	1	1	-	-	2
CO-5	3	1	1	1	2	1	1	-	1	1	-	1
CO-6	3	3	2	2	1	1	2	-	2	1	-	1
CO-7	3	3	3	3	-	2	3	-	2	2	1	2
CO-8	2	2	2	1	-	2	3	2	2	3	2	2

GUIDELINES FOR TEACHERS

To implement Outcome Based Education (OBE) knowledge level and skill set of the students should be enhanced. Teachers should take a major responsibility for the proper implementation of OBE. Some of the responsibilities (not limited to) for the teachers in OBE system may be as follows:

- Within reasonable constraint, they should manoeuvre time to the best advantage of all students.
- They should assess the students only upon certain defined criterion without considering any other potential ineligibility to discriminate them.
- They should try to grow the learning abilities of the students to a certain level before they leave the institute.
- They should try to ensure that all the students are equipped with the quality knowledge as well as competence after they finish their education.
- They should always encourage the students to develop their ultimate performance capabilities.
- They should facilitate and encourage group work and team work to consolidate newer approach.
- They should follow Blooms taxonomy in every part of the assessment.

Bloom's Taxonomy

Level	Teacher should Check	Student should be able to	Possible Mode of Assessment
Create	Students ability to create	Design or Create	Mini project
Evaluate	Students ability to justify	Argue or Defend	Assignment
Analyse	Students ability to distinguish	Differentiate or Distinguish	Project/Lab Methodology
Apply	Students ability to use information	Operate or Demonstrate	Technical Presentation/ Demonstration
Understand	Students ability to explain the ideas	Explain or Classify	Presentation/Seminar
Remember	Students ability to recall (or remember)	Define or Recall	Quiz

GUIDELINES FOR STUDENTS

Students should take equal responsibility for implementing the OBE. Some of the responsibilities (not limited to) for the students in OBE system are as follows:

- Students should be well aware of each UO before the start of a unit in each and every course.
- Students should be well aware of each CO before the start of the course.
- Students should be well aware of each PO before the start of the programme.
- Students should think critically and reasonably with proper reflection and action.
- Learning of the students should be connected and integrated with practical and real life consequences.
- Students should be well aware of their competency at every level of OBE.

ABBREVIATIONS AND SYMBOLS

List of Abbreviations

General Terms			
Abbreviations	**Full form**	**Abbreviations**	**Full form**
AAC	Autoclave Aerated Concrete	LVDT	Linear Variable Differential Transformer
BCC	Body Centred Cubic	LWAC	Light Weight Aggregate Concrete
BIS	Bureau Of Indian Standards	LWC	Light-Weight Concrete
C_2S	Dicalcium Silicate	M Sand	Manufactured Sand
C_3A	Tricalcium Aluminate	NDT	Non-Destructive Testing
C_3S	Tricalcium Silicate	OPC	Ordinary Portland Cement
C_4AF	Tetra Calcium Alumino Ferrite	PCC	Plain Cement Concrete
CPVC	Critical Pigment Volume Concentration	PIC	Polymer Impregnated Concrete
C-S-H	Calcium Silicate Hydrate	PMB	Polymer Modified Bitumen
CTD	Cold Twisted Deformed	PPC	Portland Pozzolana Cement
CTOD	Crack Tip Opening Displacement	PSC	Portland Slag Cement
EPS	Expanded Polystyrene	PVC	Pigment Volume Concentration
FCC	Face Centred Cubic	RCC	Reinforced Cement Concrete
FRC	Fibre Reinforced Concrete	SBR	Styrene-Butadiene Rubber
GFRC	Glass Fibre Reinforced Concrete	SDG	Sustainable Development Goal
HCP	Hexagonal Close Packed	SFRC	Steel Fibre Reinforced Concrete
HPC	High Performance Concrete	SINTAP	Structural Integrity Assessment Procedure
HPL	High Pressure Laminate	SSD	Saturated Surface Dry
HYSD	High Yield Strength Deformed	TMT	Thermos-Mechanically Treated
IS	Indian Standard	TMT-HCR	Thermo-Mechanically Treated High-Strength Corrosion Resistant
LCA	Life Cycle Assessment	UPV	Ultrasonic Pulse Velocity
LCC	Life Cycle Cost	UTM	Universal Testing Machine
LPL	Low Pressure Laminate		

Units used			
Abbreviations	**Full form**	**Abbreviations**	**Full form**

Cal	Calorie	l	Litre
cm	Centimetre	m	Metre
dB	Decibel	mg	Milligram
g	Gram	ml	Millilitre
h	Hour	mm	Milli metre
J	Joules	MPa	Megapascals
$kCal$	Kilo calorie	μm	Micro metre
kg	Kilogram	μPa	Micro Pascals
kN	Kilo Newton	N	Newton
kPa	Kilo Pascal	ppm	Parts per million

List of Symbols

Symbols	Description	Symbols	Description
$°C$	Temperature in degree Celsius	K	Bulk modulus
a	Crack length	K_c, K_{IC}	Fracture toughness or critical stress intensity factor
A	Area of cross-section	L	Length of sample
D, d	Diameter	L_0	Gauge length
δ	Displacement/deflection	M	Bending moment
Δl	Change in length	ν	Poisson's ratio
E	Modulus of elasticity	P	Applied load
ϵ	Strain	π	Pi
f_f	Fracture stress	R	Radius of curvature
F	Axial force	S_0	Original cross-sectional area
G	Rigidity modulus	σ	Stress
γ_s	Surface energy density	T	Torque
I	Moment of Inertia	τ	Shear stress
J	Polar moment of Inertia	τ_c	Critical resolved shear stress

LIST OF FIGURES

Unit 1 Introduction to Engineering Materials

Unit 2 Emerging Civil Engineering Materials

Unit 3 Introduction to Material Testing

Unit 4 Standard Testing & Evaluation Procedures

LIST OF TABLES

Unit 1 Introduction to Engineering Materials

Unit 2 Emerging Civil Engineering Materials

Unit 4 Standard Testing & Evaluation Procedures

CONTENTS

1 Introduction to Engineering Materials

UNIT SPECIFICS

Through this unit, we have discussed the following aspects:

- *Numerous materials utilized for application in civil engineering;*
- *Physical attributes of typical structural and geotechnical building materials being measured;*
- *Observing various types of material behaviour, including modes of failure;*
- *Process of preparing manufacturing materials, along with directions for modifications for desired material properties*

The different types of civil engineering materials are discussed to foster an appreciation towards alternative building materials and the various properties exhibited by them, to encourage the efficient use of civil engineering materials.

Along with a large number of multiple-choice questions and questions with short and long answers, the unit also includes a list of references and suggested readings that one can use as practice and grow their knowledge on civil engineering materials.

RATIONALE

This fundamental unit on engineering materials helps students to familiarize with the basic civil engineering materials. The unit highlights several popular civil engineering materials such as timber, glass, cement and steel. The unit explains the different engineering materials, their manufacturing process, testing procedures, and interpretation of test results to know whether a material meets the required standards for construction according to codal provisions. The discussion is followed by some problems related to civil engineering materials, which will further help in developing a clear idea of the concerned topics on the subject.

A good grasp of building materials is necessary for the construction of buildings and other structures. Without this information, it will be impossible to construct buildings, structures, and dwellings that are secure, cost-effective, and long-lasting, while accounting for various needs. An overview of all the civil engineering materials will help civil engineers and other stakeholders to identify suitable materials, test desired properties, and identify the suitable material for effective use.

PRE-REQUISITES

Chemistry: Chemical equations and commonly used notations (Class XII)

Physics: Mechanics (Class XII).

UNIT OUTCOMES

As outcome of this unit, one should be able to:

U1-O1: Understand engineering usage of various materials.

U1-O2: Study the ingredients and composition of various materials.

U1-O3: Comprehend the manufacturing process of various materials

U1-O4: Evaluate the physical and mechanical properties of various materials.

U1-O5: Apply appropriate material for engineering purposes.

Unit-1 Outcomes	EXPECTED MAPPING WITH COURSE OUTCOMES (1- Weak Correlation; 2- Medium correlation; 3- Strong Correlation)							
	CO-1	CO-2	CO-3	CO-4	CO-5	CO-6	CO-7	CO-8
U1-O1	2	-	2	2	-	3	2	2
U1-O2	1	-	1	-	-	-	2	2
U1-O3	-	-	-	-	-	-	1	1
U1-O4	3	3	3	3	3	3	3	3
U1-O5	3	1	2	2	2	2	3	2

1.1 INTRODUCTION

Engineering materials are a set of materials that are employed in the creation of components and structures that are man-made. Examples of engineering materials include bricks, steel bars, concrete, etc. The prime objective of an engineering material is to sustain the applied loads without failure. It should be noted that, in the field of civil engineering, failure is not limited to the maximum load-carrying capacity of the material. Rather, the engineering material should also satisfy the specific requirements required during its service life. The specific requirements can be deflection within permissible limits, fire safety, colour, etc.

In civil engineering, based on the desired application of engineering materials (i.e., applied loading and specific requirements), different engineering materials are used. For example, in a typical construction, concrete is used for low-cost load-bearing applications and glass is used for aesthetic purposes and non-load-bearing purposes. Since engineering materials have different properties, they are suitable for different applications. As an engineer, one should have a clear understanding of the different engineering materials so as to suitably utilise them in real-life applications.

The engineering materials can either be naturally occurring, like wood, or manufactured, like concrete. Traditionally, manufactured materials are classified into four categories, i.e., ceramics, metals, polymers, and composites, as shown in Fig. 1.1. In addition to the four categories, several other engineering materials also exist, like aggregates, bitumen, and adhesives.

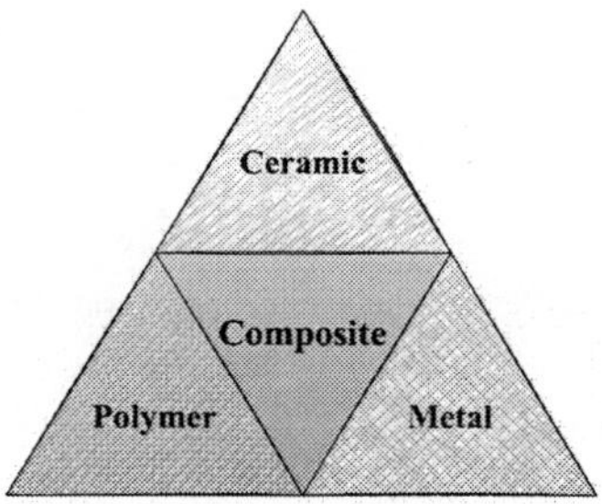

Fig. 1.1: Traditional classification of different engineering materials

Ceramics are the class of materials that are prepared by the hardening of raw materials, typically non-metallic. Commonly used ceramics include cement and clay.

Metals are the class of materials that are formed by combining one or more metals with metallic and non-metallic compounds in the form of alloys. Commonly used metals are iron, aluminium, and copper alloys. While it is also possible to directly use a metal without forming alloys, they are typically not used in civil engineering due to the weaker mechanical characteristics of pure metal.

Polymers are the class of materials that are formed by a long chain of repeating units. Commonly used polymers include plastics and adhesives.

Composites are a class of materials that are prepared by the combination of two or more materials with different proportions. Commonly used composites include concrete and paints.

In this chapter, we will cover a wide range of engineering materials commonly used in construction. This includes ceramics (cement, glass), metals (steel), composites (concrete, paints) and other engineering materials (aggregate, timber). It should be remembered that while the traditional classification system is good for developing a basic understanding, it should not be used as the basis for understanding engineering materials. To understand this limitation, let us consider the case of cement. Cement, despite being a ceramic, has more applications in the form of composites like concrete. Furthermore, its properties are widely different from other ceramic materials like glass. Therefore, to simplify the understanding, the engineering materials are discussed based on their common properties and not their traditional classification system.

1.2 CEMENT

Cement is one of the most versatile engineering materials used in construction. When mixed with water, cement becomes a paste-like material which can be given a desired shape. The cement paste then hardens over time to become a strong and durable ceramic material. The unique ability of being mouldable and hardening over time allows cement to be used in a wide range of applications like concrete making, grouting, plastering, etc. Furthermore, the relatively lower cost, ample availability, and ease of use have resulted in cement being one of the most commonly used construction materials.

1.2.1 Ingredients of cement

The raw ingredients of cement consist of Ca-rich (calcareous) and Si-rich (argillaceous) compounds. The calcareous compounds include calcium, lime, chalk, and marine shells, whereas the argillaceous compounds include clay, marl, and silica. Calcium, from calcareous compounds, is the primary source of bonding properties in cement. The typical composition of the raw ingredients of cement, their chemical formula, and their approximate weightage in the composition are tabulated in Table 1.1.

Lime: Lime gives strength to the cement. If there is a higher quantity of lime, the cement hardens quickly and gives more strength at a relatively early age, and vice versa. If there is an excess quantity of lime, some of the lime will not form cement and remains as free lime. Free lime expands in presence of water and can make the cement unsound, i.e., cement disintegrates from internal expansion of lime. Free lime also delays the setting of cement. If there is a very limited quantity of lime, strength gaining compounds will not be formed and strength will be reduced.

Silica: Silica is also responsible for giving strength to the cement. However, silica reacts much slower than lime and gives strength at a later age of concrete. If more silica is present, the cement's strength will be increased, but the cement's setting time will also be delayed; and vice versa.

Alumina: Alumina imparts the setting property to the cement, i.e., stiffening of cement paste at a very early age (typically less than 10 h). Unlike lime and silica, alumina reacts rapidly but does not contribute to the strength of the cement. If a higher quantity of alumina is used, the cement will set more quickly and will have a lower strength.

Iron oxide: Iron oxide is mainly responsible for imparting colour to the cement. Similar to alumina, iron oxide also reacts quickly and is responsible for the strength and hardness at early ages. Both iron oxide and alumina also act as a flux and lower the temperature requirement for cement manufacturing.

Calcium Sulphate: Calcium sulphate is generally added in the form of gypsum [$CaSO_4.2H_2O$]. Calcium sulphate regulates (delays) the reaction of alumina and helps in increasing the initial setting time of the cement.

Magnesia: Magnesia imparts a yellow colour to the cement. Magnesia also imparts strength and hardness to the cement. Magnesia behaves similar to lime and makes the cement unsound, if present in excess quantity.

Alkali: Alkali are not added intentionally, they become part of the cement along with other raw ingredients. Alkali, in cement, is responsible for staining and efflorescence in the structure. Alkali also undergoes expansive reaction with aggregates, thereby leading to cracking in the structure.

Table 1.1: Ingredients of Cement

Ingredients	Chemical Formula	Percentage (%)
Lime	CaO	60-67
Silica	SiO_2	17-25
Alumina	Al_2O_3	3-8
Calcium sulphate	$CaSO_4$	3-4
Iron oxide	Fe_2O_3	0.5-6
Magnesia	MgO	1-4
Alkalis	Na_2O, K_2O	0.2-1.3

In addition to the discussed chemical compounds, other chemicals like TiO_2, S and $CaCO_3$ are often added in cement raw ingredients to prepare special types of cement.

1.2.2 Manufacturing of cement

Manufacturing of the cement is carried out in three distinct operations, i.e., (i) mixing of raw ingredients, (ii) burning, and (iii) grinding. On the basis of presence of water or moisture during the mixing and burning operations, two different types of processes are used for the manufacturing of the cement;

(i) Dry process (new method)

(ii) Wet process (old method)

Dry process:

When the raw materials are very hard, the dry method is used. The dry process is expensive and cumbersome as compared to the wet process. In the dry process, calcareous and argillaceous materials are individually ground to a fine powder and then combined in predetermined proportions to form a raw meal. Raw meal is then thoroughly mixed (blending) by agitating with the help of compressed air. After blending, the raw meal is loaded into the rotary kiln for burning operation.

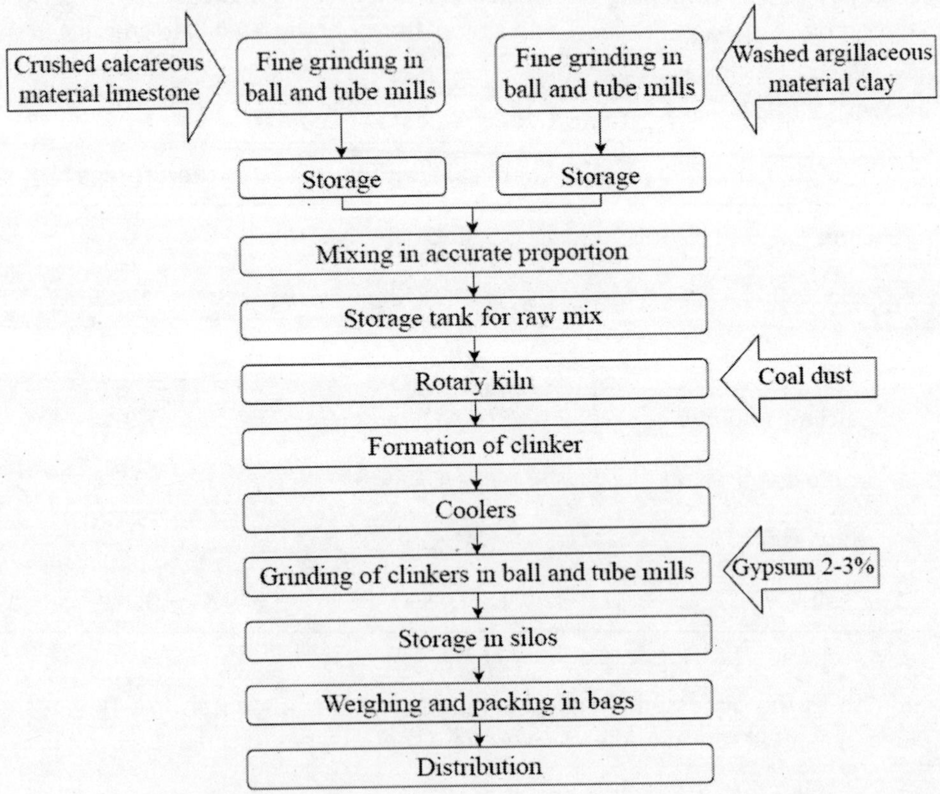

Fig. 1.2: Flow diagram of cement manufacturing – Dry process

Alternatively, a small amount of water (12% - 15%) is added with the raw meal to prepare a thick paste. The thick paste is then thoroughly mixed and used to form pellets or nodules, which is then dried and loaded into the rotary kiln. This alternate strategy of adding small amounts of water, is termed as a semi-dry process. In both dry and semi-dry processes, all operations other than blending remain the same.

In the rotary kiln, burning or calcination of raw meal is carried out at a temperature of roughly 1400°C –1500°C. In the rotary kiln, ferric oxide functions as a flux because it has a lower melting point than the other oxides. At high temperatures, the raw ingredients fuse together to form a clinker. The clinker is then cooled before grinding. To prevent the long exposure of the clinker from environmental factors like moisture and carbon dioxide, the clinker is cooled quickly. After grinding, the clinker is mixed with additional compounds, like gypsum, to form the final form of cement. The final product or cement is typically packaged in 50 kg bags, and sent for distribution. Fig. 1.2 depicts a flow diagram for the dry process of cement manufacturing.

Wet process:

In the wet process, first the raw ingredients are powdered using a ball mill. Now instead of mixing the powdered ingredients, as in the case of dry process, the powdered ingredients are used to prepare slurry. Slurry for different ingredients is prepared separately by adding water and stored in silos (storage tanks). The different slurries are then mixed in the desired proportion of raw ingredients, to produce the raw meal in slurry form. In the wet process a large amount of water is present in the raw meal, which can typically vary between (20%-40%). The raw meal is then fed into the rotary kiln for burning operations at 1500 °C –1600 °C. In the rotary kiln, the slurry loses its moisture and results in the formation of lumps. At high temperatures, the different raw ingredients in the nodule fuse together to form clinkers. It should be noted that both mixing and burning in the wet process are carried out in slurry form as compared to the dry and semi-dry process. The subsequent operation of cement manufacturing remains the same in the wet process, i.e., cooling of clinkers, grinding of clinkers, mixing with gypsum, packaging and distribution. Fig. 1.3 depicts a flow diagram for the wet process used to manufacture cement.

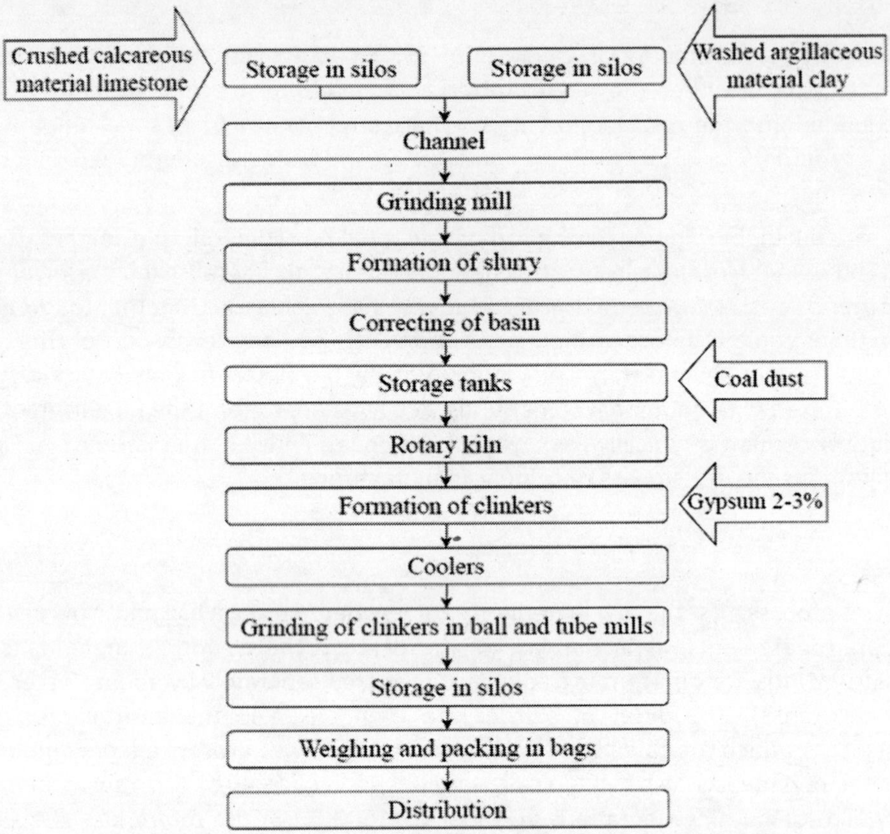

Fig. 1.3: Flow diagram of cement manufacturing – wet process

1.2.3 Composition of cement

When the raw ingredients are mixed and búrnt to form the cements, different chemicals fuse with each other to form complex compounds or minerals, referred to as Bogues compounds. Primarily there are four types of Bogues compounds, i.e., tricalcium silicate (C_3S), dicalcium silicate (C_2S), tricalcium aluminate (C_3A) and tetra calcium alumino ferrite (C_4AF).

Tricalcium silicate [$3CaO.SiO_2$]: Tricalcium silicate is known as alite, and is denoted by C_3S. C_3S makes up between 25% and 50% of cement. C_3S is primarily responsible for the 7-day strength and hardness of cement. C_3S makes the clinker easier to grind, increases resistance to freezing and thawing, hydrates quickly, produces a lot of heat, and promotes the early development of hardness and strength. Due to these advantages, C_3S is often considered as the best cementing material. However, excess C_3S content increases the hydration (chemical reactions) of cement at early ages and releases a huge amount of heat, which can lead to cracking.

Dicalcium silicate [$2CaO.SiO_2$]: Dicalcium silicate is known as belite, and is denoted by C_2S. C_2S makes up about 25-40% of cement. C_2S takes a long time for strength development. In other words, C_2S has little impact on strength and hardness at early ages or less than a month, and may take up to a year before exhibiting most of its strength. C_2S has the lowest heat of hydration which, coupled with slow hydration, lowers the heat release and lowers the risk of cracking. However, the slow hydration also makes the cement weak against freeze and thaw. C_2S has less Ca/Si as compared to C_3S, and thus provides strength to cement at a much lower Ca content. The low Ca content in C_2S makes it better against chemical and acid attacks.

Tricalcium aluminate [$3CaO.Al_2O_3$]: Tricalcium aluminate is known as celite, and is denoted by C_3A. C_3A undergoes most of the hydration within 24 hours of the addition of water into cement. The initial reaction product C_3A also releases the maximum heat during hydration. The sudden release in heat causes loss of water which leads to development of the cracks in cement. The rapid hydration of C_3A is also responsible for flash (instant) setting of the cement. The reaction products of C_3A have a tendency to expand against the attack of sulphur and form cracks, thereby making cement weak against sulphur attack.

Tetra calcium alumino ferrite [$4CaO.Al_2O3.Fe_2O_3$]: Tetra calcium alumino ferrite is known as felite, and is denoted by C_4AF. C_4AF undergoes hydration within 24 hours of the addition of water into the cement; and contributes to the initial setting of cement. C_4AF is observed to have the worst cementitious properties among all the Bogues compounds. Similar to C_3A, presence of C_4AF makes cement prone to flash setting and attack of the sulphur.

Each of the Bogues compounds have a different contribution to strength (i.e. cementation property) and cementation behaviour (i.e. heat of hydration and rate of hydration). Based on the difference in properties, Bogues compounds can be ranked as follows:

Heat of hydration: $C_3A > C_3S > C_4AF > C_2S$

Rate of hydration: $C_4AF > C_3A > C_3S > C_2S$

Cementitious property: $C_3S > C_2S > C_3A > C_4AF$

Table 1.2: Heat of hydration for typical Bogues compounds

Bogues compounds	**3 days (Cal/g)**	**90 days (Cal/g)**
C_3A	210	310
C_4AF	70	100
C_3S	60	105
C_2S	10	40

1.2.4 Types of cement

Ordinary Portland cement:

Ordinary Portland cement (OPC) is the simplest and one of the most commonly used forms of cement. OPC essentially consists of clinker and gypsum. OPC may be considered as the basic form of cement, as other types of cement are typically compared with OPC only. For example, rapid hardening cement shows strength faster than OPC.

Portland pozzolana cement:

Portland pozzolana cement (PPC) is prepared by inter-grinding cement clinkers with up to 35% pozzolanic material. Pozzolanic material is essentially a siliceous or aluminium compound which does not possess any binding property but, in presence of water, reacts with lime released during the hydration of cement and leads to the formation of cementitious compound. Ex. 10surkhi, fly ash, rice husk ash. PPC offers a slower rate of gain of compressive strength. PPC has higher plasticity, low heat of hydration, low cost, higher resistance against chloride and sulphates, and higher resistance against volume changes.

$$\underset{}{Ca(OH)_2} + \underset{\text{(Non-cementitious compound)}}{\text{pozzolana}} + H_2O \rightarrow \underset{\text{(Cementitious compound)}}{\text{C-S-H}} \quad (1.1)$$

Quick setting Portland cement:

Quick setting Portland cement is prepared by reducing the gypsum content and adding a small amount of aluminium sulphate to get the quick setting property. Quick setting Portland cement is much finer than OPC. This cement sets quickly, while the rate of strength gain is similar to OPC. Typically, the initial setting and final setting time are 5 minutes and 30 minutes, respectively. Quick setting Portland cement is commonly used when concrete is to be laid under water.

Portland slag cement:

Portland slag cement (PSC) is prepared by integrating granulated blast furnace slag (upto 65%) with cement clinker and gypsum. PSC, among the commercially used cements, requires the lowest amount of clinker and utilizes the highest amount of industrial waste (i.e., slag). PSC offers low heat of hydration, lower cost and higher resistance against the attack of chloride and sulphate. PSC is commonly used for mass concreting and water retaining structures.

Rapid hardening cement:

Rapid hardening cement is prepared by finely grinding cement clinker (specific area being greater than 3250 cm^2/g). This cement increases strength more quickly. The strength of rapid hardening cement at the age of 3 days is about the same as that of OPC in 7 days. Rapid hardening cement should not be confused with quick-setting cement, as the former is about over time strength gain and later is about initial strength gain or setting. Rapid hardening cement finds its application in cold weather concreting, pavement construction, prefabricated structures and emergency repair work.

Extra rapid hardening cement:

Extra rapid hardening cement is obtained by mixing a small quantity of $CaCl_2$ with rapid hardening cement. Addition of $CaCl_2$ further accelerates setting and rate of strength gain in rapid hardening cement. This cement is ideal for concreting in cold weather because it sets, hardens and releases heat more quickly in the early stages of hydration. At an age of 2 days, extra rapid hardening cement has typically 25% higher strength than rapid hardening cement. However, after a long period of time, the strength of extra rapid hardening cement and rapid hardening cement are nearly the same. The application of this cement is the same as rapid hardening cement, but the use of extra rapid hardening cement is prohibited in prestressed concrete construction. Maximum time of using this cement is 20 minutes which includes mixing, transporting, placing and compaction.

Sulphate resisting cement:

Sulphate resistant cement is prepared by reducing the percentage of lime and alumina in such a way that the proportion of C_3A is not more than 5% and the combined proportion of C_3A and C_4AF is not more than 25%. During civil engineering applications, cement is often exposed to sulphur in the form of sulphates, especially magnesium sulphate. Sulphate predominantly attacks the hydrated form of C_3A and results in the formation of calcium sulphoaluminate, which is an expansive reaction. The increase in volume from sulphate attack, causes internal expansion/swelling and leads to the development of cracks in cement paste. The lower quantities of C_3A and C_4AF in sulphate resisting cement, reduces the risks associated with sulphate attack. This type of cement finds its application in foundation work, sewage treatment work, marine construction and concrete pipes to be laid in marshy areas.

Super sulphated cement:

Super sulphated cement is prepared by intergrading 80-85% granulated blast furnace slag, 10-15% gypsum and 5% clinkers. It is the type of cement in which resistance against the sulphate attack is even more than that of sulphate resisting cement. In super sulphated cement heat of hydration and strength is less than OPC. The application of this cement is the same as that of sulphate resisting cement.

Hydrophobic cement:

Hydrophobic cement is prepared by mixing a small dose of water-repellent film-forming material, like stearic acid, with OPC in a grinding process. The water-repellent layer prevents the movement of water and water-based impurities which damage the cement. Hydrophobic cement has a better water resistance and water im-permeability, resulting in better durability characteristics. The strength characteristics of hydrophobic cement are almost identical to those of OPC.

IRS 40 cement:

IRS 40 cement is a special type of patented cement used by the Indian railway for the manufacturing of railway sleepers. This cement is produced by finely grinding the cement clinkers and increasing the percentage of C_3S.

High alumina cement:

High alumina cement is prepared by significantly increasing the C_3A content, using high alumina raw materials like bauxite, as compared to ordinary Portland cement. The ratio of alumina to lime, measured in terms of weight, must be between 0.85 and 1.30, and the total alumina content must not be less than 32%. High C_3A content allows the cement to set very quickly. High alumina cement quickly increases the ultimate strength, around 40 N/mm^2 after 1 day and 50 N/mm^2 after 3 days. It is characterized by its dark colour, high early strength, high heat of hydration and resistance to freezing and thawing. The initial and final setting time of high alumina cement is 3.5 hours and 5 hours, respectively. Instead of the calcium silicates found in Portland cement, monocalcium aluminate (Al_2O_3, CaO) makes up most of the compound, giving it quick hardening qualities. Furthermore, high alumina cement can withstand high temperatures and can be used as a refractory cement.

1.3 AGGREGATES

Aggregates are engineering materials which are used as fillers for a wide range of engineering applications. Aggregates can be used for applications ranging from preparation of concrete, sub base for pavements, ballast for railways. In general aggregates are naturally occurring stones or sand, with good mechanical and physical characteristics. Aggregates can also be of artificial or waste origin, provided they satisfy the desired application requirements. Aggregates are typically of lower cost than most engineering materials and hence widely used for filler applications.

1.3.1 Classification of aggregates

Aggregates are mainly classified on the basis of size or gradation of particles with reference to IS 383. On the basis of gradation, aggregates can be of three types:

1. Fine aggregates: Aggregates for which most of the material passes through a 4.75 mm IS sieve are known as fine aggregates. Fine aggregates are commonly known as sand. The fine aggregates are further divided in four zones on the basis of gradation, as shown in Table 1.3. As one moves from Zone I to Zone IV, the fine aggregate becomes finer. If any fine aggregate fails to satisfy the specified grading zones, the percentage of aggregate passing 600 μm sieve is typically used to classify the fine aggregates. For most construction applications Zone II fine aggregates are preferred.

Table 1.3: Classification of fine aggregates

S. No.	IS Sieve	Percentage passing			
		Grading Zone I	**Grading Zone II**	**Grading Zone III**	**Grading Zone IV**
1	10 mm	100	100	100	100
2	4.75 mm	90-100	90-100	90-100	95-100
3	2.36 mm	60-95	75-100	85-100	95-100
4	1.18 mm	30-70	55-90	75-100	90-100
5	600 μm	15-34	35-59	60-79	80-100
6	300 μm	5-20	8-30	12-40	15-50
7	150 μm	0-10	0-10	0-10	0-15

2. Coarse aggregates: Aggregates for which most of the material is retained on a 4.75 mm IS Sieve are known as coarse aggregates. Coarse aggregates are commonly known as gravel or aggregates. The coarse aggregates are further divided on the basis of maximum size of aggregate and the gradation of aggregates, as shown in Table 1.4 and Table 1.5. Aggregates which dominantly consist of a single size range are known as *single-sized aggregates* (Table 1.4). On the other-hand well graded aggregates comprising all sizes of aggregates are known as *graded aggregates* (Table 1.5).

Table 1.4: Classification of single-sized coarse aggregates

S. No.	IS Sieve	Percentage passing for single-sized aggregate of nominal size					
		63 mm	**40 mm**	**20 mm**	**16 mm**	**12.5 mm**	**10 mm**
1	80 mm	100	-	-	-	-	-
2	63 mm	85-100	100	-	-	-	-
3	40 mm	0-30	85-100	100	-	-	-
4	20 mm	0-5	0-20	85-100	100	-	-
5	16 mm	-	-	-	85-100	100	-
6	12.5 mm	-	-	-	-	85-100	100
7	10 mm	0-5	0-5	0-20	0-30	0-45	85-100
8	4.75 mm	-	-	0-5	0-5	0-5	0-20
9	2.36 mm	-	-	-	-	-	0-5

Table 1.5: Classification of graded coarse aggregates

S. No.	IS Sieve	Percentage passing for graded aggregate of nominal size			
		40 mm	20 mm	16 mm	12.5 mm
1	80 mm	100	-	-	-
2	63 mm	-	-	-	-
3	40 mm	90-100	100	-	-
4	20 mm	30-70	90-100	100	100
5	16 mm	-	-	90-100	-
6	12.5 mm	-	-	-	90-100
7	10 mm	10-35	25-55	30-70	40-85
8	4.75 mm	0-5	0.-10	0-10	0-10
9	2.36 mm	-	-	-	-

3. All in aggregates: Aggregates for which the material consists of both coarse and fine aggregates in a well distributed range are known as all in aggregates. For most engineering applications, all in aggregates are required, which are prepared by mixing of fine and coarse aggregates in a predetermined ratio. If all in aggregates are available, they may be directly used for engineering applications. IS 383 has defined the all-in-aggregate grading for a maximum nominal size of 40 mm and 20 mm only. For all other nominal size of all-in-aggregates, the grading may be defined as per the engineering application.

In addition to classification on the basis of particle size distribution, the aggregate may also be classified on the basis of their origin (e.g., natural and artificial aggregates) and characteristics (e.g., dense and light-weight aggregates). However, these are not very strict forms of classification, and thus not discussed in detail.

1.3.2 Manufactured sand (M sand)

M sand or manufactured sand is the fine remain obtained after crushing of rocks or granite. It is used as fine aggregates for preparing mortar or concrete for construction applications. M sand is different from naturally occurring river sand in terms of physical and mineralogical characteristics, and exhibits several advantages when used in construction. M sand has gained a lot of significance, as the has been greatly promoted as the substitute of natural river sand.

Advantages of M sand:

- Improved quality of concrete: M sand aids in producing higher-quality concrete since it has less impurities like dust, clay, and other contaminants. Additionally, the use of M sand overcomes the structural defects in concrete such as presence of capillarity, segregation, bleeding, and honeycombing.

- Durability of concrete: M sand is produced from a particular type of granite, and has the ideal physical and chemical characteristics for concrete construction. M sand helps the concrete structures to bear extreme weather conditions and provide corrosion resistance.
- Eco-friendly: By using M sand, it is no longer necessary to dredge riverbeds to extract river sand, which can have detrimental environmental effects such as water depletion and effect on the safety of dams and bridges.

Disadvantages of M sand:

- The flaky or angular particles produced by improper crushing may require the addition of more water and cement. The higher cement requirement for M sand concrete may not be cost-effective for pumpable concrete in high-rise constructions.
- Due to the greater manufacturing prices as compared to river sand, M sand may not be economical over natural river sand in some regions of the country.

Applications of M sand:

M sand can be used almost everywhere as a substitute for river sand. Some of the well-recognized applications in Indian Standards are as follows:

- Concrete: IS 383 (2016) allows the use of M Sand for all concreting applications except for prestressed concrete. The standard allows for a higher percentage of fine dust and percentage of particles passing 150 microns in M sand as compared to natural sand.
- Concrete using recycled fine aggregates: IS 383 (2016) extends the definition of M sand to include recycled fine aggregates also. However, the extent of utilization of recycled fine aggregates is limited to lean concrete and partial substitution of river sand for all other types of concrete. Furthermore, use of recycled fine aggregates is slightly more relaxed than other manufactured aggregates.
- Plaster: IS 1542 (1992) allows for the use of M sand with higher percentage of fine dust and percentage of particles passing 150 microns as compared to natural sand for the plastering applications.
- Masonry mortar: IS 2116 (1980) allows for the use of M sand with higher percentage of fine dust as compared to natural sand for the preparation of masonry mortar.

1.4 WATER

Water is the largest consumed natural material on earth. It is used in a large number of engineering applications ranging from concrete casting, curing, compaction of soil, to varnish. The quality of water plays a major role in its application, especially in case of concrete. For both mixing and curing operations of concrete, the water should be free from injurious amounts of impurities like salts, acids, alkalis, oils, and organic materials. Ideally, distilled water should be used for engineering applications. However, considering the cost of distillation, potable water is preferred for most engineering applications.

According to IS 456, following points should be considered while using water of any source:

1. Average compressive strength of concrete prepared by a given source of water should not be less than 90% as compared to concrete prepared by distilled water.
2. Initial setting time from the given source of water should not be less than 30 min and shall not differ by $\pm$ 30 min as compared to the initial setting time when distilled water is used.
3. In terms of acidity, pH of water shall not be less than 6. Also, to neutralize 100 ml sample of given water, not more than 5 ml of 0.02 N NaOH is to be used.
4. Similarly, in terms of basicity, to neutralize 100 ml sample of given water, not more than 25 ml of 0.02 N H_2SO_4 is to be used.
5. Sea water should be avoided for reinforced cement concrete to prevent the risk of corrosion, which arises from the presence of salts.
6. The solids in given source of water should be limited to the values as listed in Table 1.6. Also, tannic acid and iron compounds should be absent from the water.

Table 1.6: Permissible limit for solids in water used for concrete

S. No	Type of solid	Permissible limit
1	Organic	200 mg/l
2	Inorganic	3000 mg/l
3	Sulphates (SO_3)	400 mg/l
4	Chlorides (Cl)	2000 mg/l (for plain cement concrete) 500 mg/l (for reinforced cement concrete)
5	Suspended matter	2000 mg/l

1.5 CONCRETE

Concrete is one of the most widely used building material in the construction industry. Concrete is a composite material typically consisting of binding materials, fine aggregates, coarse aggregate and water, with possible additions of fibres and admixtures. In general, cement is used as the binding material with or without the presence of mineral admixtures; however other binders like lime and geopolymer may also be used. The freshly mixed concrete is called wet or green concrete. Green concrete has a plastic fluid like nature, which allows for moulding the composite material to desired shapes. Over time green concrete loses its fluidity and sets to become a solid mass called hardened concrete. Hardened concrete is strong and durable, which is a desirable quality in building materials.

Mouldability, early hardening, and high compressive strength are the main factors contributing to the widespread use of concrete. Historically, the structures were built using concrete with lime as the binding material. Since the invention of cement, use of lime has become limited and cement is mostly used as the binding material. At present, concrete is used in various constructions, including buildings, bridges, piers, dams, weirs, and liquid retaining structures.

1.5.1 Classification of concrete

Concrete is a composite material with a large number of possible variations, ranging from type of raw materials to the method of preparation. Some of the commonly used classifications of concrete are discussed below.

Based on binding or cementitious material:

Cement, lime, gypsum, and geopolymer are the typical binding or cementitious materials used for producing concrete, and hence result in cement concrete, lime concrete, gypsum concrete and geopolymer concrete respectively.

Based on grade of a concrete:

Grade of a concrete is defined by the characteristic compressive strength of concrete blocks tested at 28 days. Concrete with grade less than or equal to 15 N/mm^2 is known as lean concrete, whereas high-strength concrete refers to a grade greater than or equal to 65 N/mm^2. The rest of the grades of concrete, i.e., 20 N/mm^2 to 60 N/mm^2, are referred to as ordinary and standard grades of concrete. In general use, lean concrete and standard grade concrete are also referred to as low strength concrete and medium strength concrete.

Based on the proportion:

The proportions of the ingredients define the properties of concrete, including the grade of concrete. For most engineering applications the proportion is set as per the required properties and is known as design mix concrete. Design mix concrete are designated with 'M' followed by the grade of concrete, e.g., M 40. In some cases, prespecified mix proportions may also be used and are known as nominal mix concrete. Nominal mix concrete is designated same as design mix concrete, e.g., M 15. The various nominal mix proportions listed in IS 456 are provided in Table 1.7. IS 456 suggests the use of design mix concrete for all applications with grade M 20 or higher.

Table 1.7: Proportions for nominal grades of concrete

Grade of concrete	Maximum quantity of dry aggregate by mass per 50 kg of cement (kg)	Proportion of fine aggregate: coarse aggregate (by mass)	Maximum quantity of water per 50 kg of cement (l)
M 5	800	Generally, 1:2, but subject to an upper limit of 1:1.5 and a lower limit of 1:2.5	60
M 7.5	625		45
M 10	480		34
M 15	330		32
M 20	250		30

Based on bulk density:

Bulk density is another important property of concrete which is used during the analysis and design of structures. The normal bulk density of concrete is defined as 2400 kg/m^3. Use of different raw materials affect the bulk density of concrete. Indian standard broadly recognizes the concrete as normal weight, light-weight and heavy concrete, without specifying the limits for classification. However, for typical applications, classification as shown in Table 1.8 may be followed.

Table 1.8: Classification based on denseness property

Classification	Bulk Density
Light weight concrete	less than 1800 kg/m^3
Normal weight concrete	1800 to 2500 kg/m^3
Heavy concrete	greater than 2500 kg/m^3

Based on casting place:

Concrete can be cast both onsite and offsite. If the concrete is cast at the site, then it is known as cast-in-situ concrete. If concrete is cast off site and then placed at the site, then it is known as precast concrete. In precast concrete, modular units are cast in a factory and transported to the site for erection.

1.5.2 Stages in the production of the concrete

The process for producing the concrete, as shown in Fig. 1.4, is briefly described as follows:

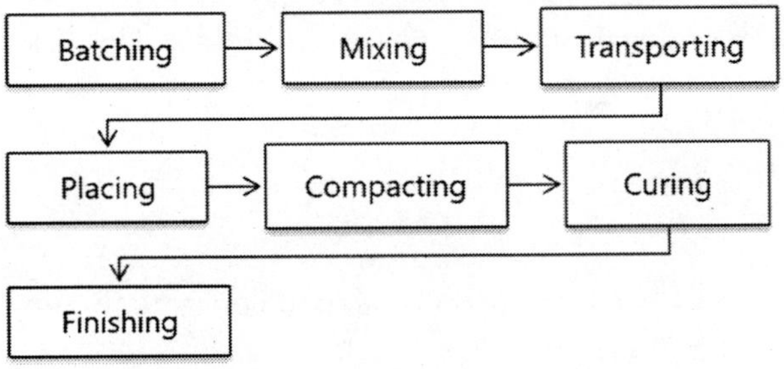

Fig. 1.4: Process of concrete production

Batching:

It is defined as measuring the quantity of ingredients for the concrete mix in the required proportion. It can be done in two ways:

(a) *Volume batching:* The quantity of each ingredient is measured in terms of volume using a box called *"farma"*. The volume of farma is generally equal to the volume of one bag of cement (0.035 m^3). This type of batching is used for small-scale projects.

(b) *Weight batching:* The quantity of the ingredient of the required proportion is taken with the respective weight of the materials. Different weight batchers are used depending upon the size of the projects, like for small projects, weighing is done manually using weighing balance whereas for big projects weighing is done with weighing buckets. This type of batching is suitable for large scale projects.

Mixing:

It is a process of combining the different ingredients of concrete to produce a homogenous mass, either manually or by using a machine called a mixer.

Transporting:

It is the process of transporting the concrete from the place of mixing to the pouring locations. The time specified for transporting the concrete after the mixing is a maximum of three hours, depending on the type of cement. Transporting should be carried out without any disturbance so that the concrete does not lose its homogeneity and consistency.

Placing:

It is the process of pouring or placing the transported concrete with maximum care so that it does not lose its consistency and homogeneity. Concrete should not be poured from a great height to avoid segregation, i.e., separation of aggregate from cement paste.

Compaction:

It is the process of improving the packing of freshly poured concrete. It is done by removing the air voids by agitating the concrete mix using appropriate compaction techniques. Typically, compaction is carried out using vibrators and rollers. Over compaction can result in segregation, i.e., separation of cement paste from aggregates, and reduce the strength of the hardened concrete.

Curing:

Cement requires water to react and form hardened concrete. However, over time the concrete loses its water due to several environmental factors, which prevents the cement from reacting. Curing is the process of providing moisture to the freshly placed concrete, to ensure sufficient water availability for cement hydration and strength gain. In other words, curing is the process of keeping the concrete in a damp state for the required duration.

Finishing:

It is the process of giving a pleasant appearance to the rough concrete surface by making it smooth and level.

1.5.3 Plain cement concrete (PCC)

The cement concrete without any reinforcement is called plain cement concrete (PCC). PCC has considerable strength in compression and comparatively very less strength in tension. In some cases, even if the reinforcement is provided, the structures or its elements are said to be PCC if the reinforcement is not considered for strength calculation. As per IS 456, concrete grades higher than M 15 can be used as plain concrete. The unit weight of PCC is taken as 2400 kg/m^3. PCC is generally used for constructing concrete pavements, pathways, footpaths, pedestals and base of footings where little or no tensile strength is required.

1.5.4 Reinforced cement concrete (RCC)

Reinforced cement concrete (RCC) is a composite material prepared by adding continuous reinforcements to the concrete. Reinforcements are typically made of steel bars, but can be made from different materials like polymer, aluminium and bamboo. Concrete is strong in compression, but weak in tension and shear. In RCC, steel reinforcement is added to resist tension and shear while the concrete resists compression. Stress transfer between steel and concrete is made possible by the bond between the steel and the surrounding concrete. Furthermore, in a typical RCC, reinforcement carry tensile loads even after the concrete has cracked under tension, and prevent the sudden failure of structural members, as shown in Fig. 1.5.

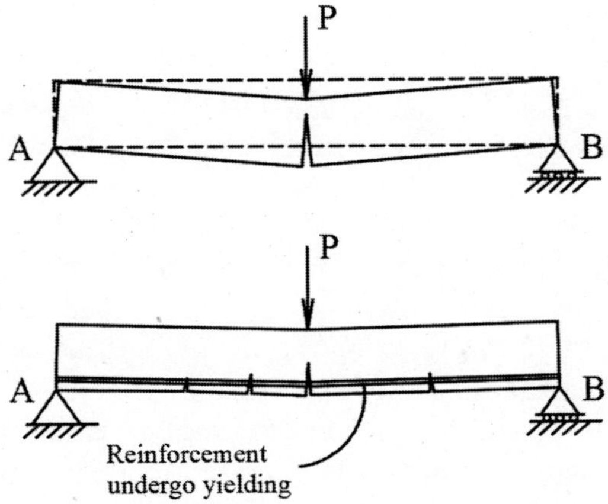

Fig. 1.5: Difference between PCC and RCC

In some cases, like columns, reinforcements are added to improve the compressive strength of concrete. In other terms, RCC can be used to build structural members that withstand high compression, shear and tensile stresses by placing reinforcement in the desired zone. Due to its adaptability, RCC is widely utilized in the construction industry from small structural components like beams and columns to enormous projects like dams and bridges. As per IS 456, the minimum concrete grade of M 20 is mandatory for RCC. The unit weight is generally taken as 2500 kg/m^3.

1.5.5 Special concretes

Concrete is a composite material and its properties can be modified by changing the composition of raw materials. Modification in composition helps in compensating for the shortcomings of concrete, like poor tensile strength, and also imparts special characteristics to concrete, like light weight. Concretes modified to have different properties for specific applications are known as special concretes. Special concretes can be produced for specific applications, such as low weight, fire protection, radiation shielding, etc. In broad sense, the special concretes are modified for the following:

- physical properties, such as density, thermal conductivity, sound conductivity, etc.
- mechanical properties, such as compressive strength, tensile strength, impact resistance, etc.
- durability properties, such as resistance from chemical attach, freeze and thaw, abrasion, etc.

In this regard, different types of special concrete commonly used in civil engineering applications are described as below:

Fibre reinforced concrete (FRC):

In FRC concrete is reinforced by the dispersion of short discontinuous fibres. Fibres are made of materials with good tensile properties like asbestos, steel, sisal, glass, carbon, polypropylene, nylon, etc. Fibres prevent the cracking of concrete under load applications, thus improving the tensile performance of concrete. The FRC is mainly used for the construction of hydraulic structures, airport pavements, roads, bridge decks, heavy-duty floors, tunnel linings, etc.

Fig. 1.6 shows the behaviour of fibre-reinforced concrete (FRC). In FRC, the yielding strain of fibres is higher than the tensile cracking strain of cement matrix. As a result, when load is applied to the FRC, the cement matrix cracks before yielding of fibres. The fibres hold the cracked section by acting as a bridge for load transfer. It should be noted that bridging action occurs only when the pull-out resistance and yielding strength of fibres are larger than the load being transferred from the fibres of the cracked section. Due to bridging action the cracked section carries additional load, and thus the fibres help in improving the tensile performance of concrete. The average binding strength between the fibres and the matrix, the number of fibres across the crack, the length and diameter (aspect ratio) of the fibres, shape and material of fibres, dispersion of fibres and orientation of fibres, all affect the tensile performance of FRC.

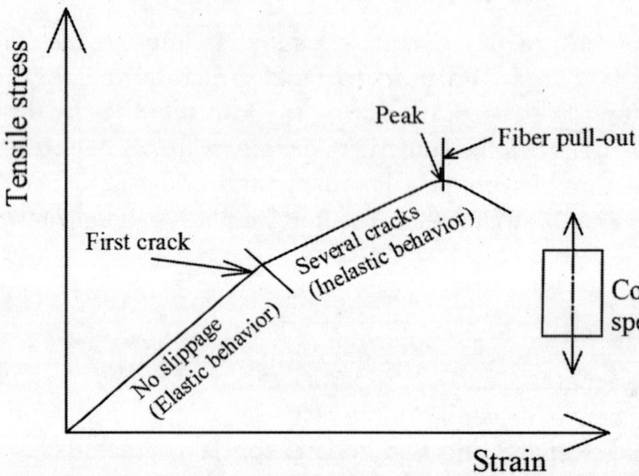

Fig. 1.6: Behaviour of FRC

The main advantages of FRC are as follows:

(i) Concrete has better crack resistance; making them excellent for water retaining structures.

(ii) Higher strain carrying capacity of fibres allow for the use of thin sections of concrete.

(iii) FRC improves ductility, impact resistance, tensile strength, and bending strength.

IS 456 permits the use of fibres in concrete, without providing any special guidelines for the use of same. One should note that addition of fibre lowers the workability of concrete. In order to compensate for the loss in workability FRC needs relatively finer aggregate and higher cement paste content than regular concrete. The preparation of a good quality FRC is also a major challenge. Addition of fibres can also result in air entertainment or meshing of fibres in form of balls. Precautions should be taken to ensure that a homogenous mixture of FRC is used for construction.

Two of the most commonly used fibres are steel fibres and glass fibre used for SFRC and GFRC respectively.

(i) Steel Fibre Reinforced Concrete (SFRC):

In SFRC, fibres made of steel are added in the concrete. The high tensile strength and ease of moulding in desired shape has made steel a popular choice for fibres. The commonly used form of fibres has a circular cross-section, diameter varying between 0.25 mm and 0.75 mm, with bent or deformed ends for better pull out resistance. Steel fibres can vary in terms of cross-section, shape, type of end bending and fibre length. When compared to ordinary concrete, the flexural strength may be improved as high as upto three times by the use of steel fibres. Following figure shows steel fibre reinforced concrete and steel fibres.

Fig. 1.7: Steel Fibre Reinforced Concrete and steel fibres
(References: Civiconcepts.com; Behbahani et al. 2011)

(ii) Glass Fibre Reinforced Concrete (GFRC):

In GFRC, high-strength glass fibres are added in the concrete. Glass fibres are lighter in weight, resistant to chemical actions like corrosion and have low thermal conductivity as compared to steel fibres. Resistance to corrosion is one of the main factors for the use of glass fibres in concrete. GFRC is commonly used for building facades, fire resistance and high-strength lightweight concrete applications. Additionally, glass fibre does not interfere with radio frequencies, while steel fibre does and are better for large structures, like multi-storey buildings.

Light-weight Concrete (LWC):

Normally concrete has a significant weight, which contributes as the load on the lower levels or support of the structure. Lower weight of concrete means lower dead load and more economical construction, especially in high-rise structures. A better construction material for such structures is LWC. LWC has significantly lower density, as low as 300 kg/m^3, as compared to normal weight concrete, having a density of 2400 kg/m^3. LWC also helps in thermal insulation and can contribute to sound insulation, making them ideal for multi-storey residential structures. It should be mentioned that hollow blocks and panels provide similar properties as LWC. However, as hollow blocks and panels are prepared by modification of shape using normal weight material, they are not treated as LWC.

LWC can be produced by either of the following:

(i) making concrete matrix cellular through the introduction of pores in cement matrix

(ii) through the use of lightweight aggregate in place of normal weight aggregates.

Let us first discuss about the cellular LWC also referred to as cellular concrete. Generally, the cellular concrete uses additives for making the concrete matrix more porous or cellular. Based on the manufacturing method or type of additive, cellular concrete can be further classified as foam concrete and aerated concrete.

Foam concrete: Foam concrete is produced by mixing concrete or mortar with a foaming agent or admixture. Alumino-sulpho-naphthene is a typical example of a foaming agents. Foaming agents result in the formation of foam cells, which solidify to create the cellular concrete. Foam concrete has low strength, about 2.5 to 7.5 N/mm^2, and thermal conductivity, about 0.10 to 0.40 kCal/h/m/°C. The strength of foam concrete can be improved by using wire mesh as a reinforcing material, resulting in strength as high as 15 N/mm^2 with thermal conductivity of 0.40-0.60 kCal/h/m/°C. Foam concrete is often preferred in the outer walls due to its insulation characteristics.

Aerated concrete: Aerated concrete or gas concrete is produced by mixing concrete or mortar with an aerating agent or admixture. Aluminium powder is typically used as the aerating agent. In presence of water aluminium reacts to release hydrogen gas (Eq. 1.2). Releases gas creates a porous shape in fresh concrete till the cement sets and forms a cellular structure in aerated concrete. In order to prevent the gas from escaping the fresh mix, it is important to synchronize the end of gas formation with the start of mix setting. In many cases autoclaving is used for the production of aerated concrete. Autoclaving simultaneously creates a favourable condition for setting and gas formation, thus producing aerated concrete. In construction, autoclave aerated concrete (AAC) blocks have emerged as one of the commonly used construction materials.

$$2Al + 3Ca(OH)_2 + 6H_2O \rightarrow 3CaO.Al_2O_3.6H_2O + 3H_2 \tag{1.2}$$

The second type of LWC is prepared by light weight aggregates, and is also known as light weight aggregate concrete (LWAC). Lightweight aggregates are typically porous or hollow coarse aggregates with significantly lower density than normal aggregates, like pumice. Lightweight aggregates can also include fine aggregates, like cenospheres, and non-porous low-density materials like rubber or thermocol (EPS) beads. As compared to cellular concrete, the pores in LWAC are concentrated at the aggregates and not the cement paste matrix.

Characteristics of LWC:

- Unit Weight: The unit weight of LWC is generally lower than 1800 kg/m^3 in an air-dried state, with density going as low as 300 kg/m^3 based on desired applications.
- Strength: The 28th-day compressive strength of LWC is lower than normal concrete. However, it should be noted that lightweight concrete does not imply low strength concrete. Higher cement dosage, lower w/c ratio and other suitable modifications are done to ensure the desired strength of concrete. Lightweight aggregates with controlled porosity have also been developed to attain a high strength of 70 to 75 MPa.

- Thermal Insulation: Thermal conductivity is typically 1/10 times the normal concrete and 1/4 times the bricks.
- Durability: Durability is reduced by the increased permeability in a typical LWC.
- Economy: Quantity of material reduces but the cost of production increases. Overall, the use of LWC becomes economical due to the cost saving from reduced dead load on structures.

Applications of LWC:

- Precast floor and roofing modules
- As non load-bearing walls
- As insulation covering for building in the form of external walls

High Performance Concrete (HPC):

HPC is a broad classification of concrete materials that exhibit one or more desirable properties, not typically observed in conventional practice. Improved placement and consolidation without segregation, long-term mechanical properties, high and early-stage gain in strength, hardness, resistance to impact, durability, service life (> 75 years) in severe environments, flowability, as well as low heat of hydration are a few examples of characteristics that define a HPC.

As there is no specific limit for the properties of a conventional material, classification of high-performance concrete often becomes subjective to interpretation. Even in case of characteristic properties like compressive strength, the minimum requirement of strength for the concrete to be classified as HPC changes for different countries. It should also be kept in mind that strength is not the only criteria of HPC, and a concrete with normal strength can also classified as HPC for better durability or flow properties. Furthermore, durability includes several factors, like permeability, de-icer scaling resistance, freeze-thaw resistance and abrasion resistance. HPC can be made any combination of these qualities are important performance criteria. Thus, one should keep in mind that HPC is a broad classification which accounts improves one or more properties in terms of strength, workability and durability, without sacrificing other properties.

High strength concrete, high workability concrete, lightweight concrete, foamed concrete, self-compacting concrete, aerated concrete, etc. are some examples of high-performance concrete.

Precise modifications are required in terms of material used and production methodology, for preparing HPC. In terms of materials HPC are designed using a wide range of mineral admixtures (fly ash, slag, silica fume, metakaolin, etc.), chemical admixture (plasticizers, viscosity modifying agents, aerating agents, etc.), fibre reinforcements (glass fibre, steel fibre, wire mesh, etc.), in addition to specialised cement and aggregates wherever required. In terms of production methodology, IS 10262 (2019) only covers guideline for high strength concrete and self-compacting concrete. For all other types of HPC, standard suggests the use of specialised literature or extensive trials. As a large number of factors are involved, HPC are highly sensitive to changes. Therefore, care should be taken when producing HPC.

Concrete polymer composite:

Concrete polymer composite is a special type of concrete which uses polymer as a partial or complete substitute of cement or lime as binding materials. Polymer as binding material has lower porosity which results in better strength and durability characteristics. Vinyl acetate copolymer emulsions, homopolymer emulsions, and polyvinyl acetate are some examples of polymer based binding materials. The primary drawback of polymer concrete is its cost. Concrete polymer composite is best suited for applications requiring higher durability, like sewage disposal operations, marine environment and construction of nuclear plants.

On the basis on production, polymer concretes can be broadly classified in four categories, as follows:

1. *Polymer impregnated concrete (PIC):* In case of PIC, hardened cement concrete is impregnated by a low viscosity monomer and polymerized. This process is typically carried out after construction of structure, and polymerization is carried out by either radiation, heat, or chemical activation. Styrene, methyl methacrylate (MMA), acrylonitrile, and t-butyl styrene are some of the monomers that are used in PIC.
2. *Polymer cement concrete (PCC):* PCC is produced by mixing monomers along with other raw ingredients of concrete. While PCC is not as effective as PIC, the process of production is much simpler for PCC. Polyester-styrene, epoxy styrene, furans, and vinylidene chloride some examples of monomers used in PCC.
3. *Polymer concrete:* In polymer concrete, rather than using cement, a polymer binder is used. Polymer concrete can have a significantly high strength of over 100 MPa, and are preferred where low dead load is required.
4. *Partially impregnated and surface coated concrete:* In partially impregnated concrete, polymer coating is focused on the exposed concrete surfaces. To produce this concrete, hardened cement concretes are soaked in a liquid monomer, such as methyl methacrylate, and then polymerized to seal the surface porosity. It may be used to improve the durability of structures where complete impregnation is not needed, like dams and bridges.

1.6 BRICKS

Brick is one of the oldest and most commonly used form of engineering material. While concrete is more versatile than bricks, bricks are popular for their simplicity and ease of application. A typical brick is a rectangular block of nominal dimensions 190 mm × 90 mm × 90 mm. The brick, in general, follows a general thumb rule for its dimensions (Eq. 1.3). However, the brick can be of other dimensions also, this includes non-modular bricks and partial dimension bricks.

$$\begin{aligned}(length\ of\ brick\ +\ thickness\ of\ mortar) \\ = 2 \times (width\ of\ brick\ +\ thickness\ of\ mortar) \\ = 2 \times (height\ of\ brick\ +\ thickness\ of\ mortar)\end{aligned} \tag{1.3}$$

1.6.1 Classification of bricks

Conventionally bricks were made from clay. However, due to growing environmental concern, there is a decrease in use of clay and greater emphasis is being given to the use of fly ash. On the basis of material used the bricks can be classified as:

1. **Burnt clay bricks:** Burnt clay bricks are prepared by clay or soil. Suitable soil is mixed with water and moulded to desired shape. The clay is then burned at a temperature greater than 1000 °C, where various minerals fuse together to form burnt clay bricks.

2. **Burnt clay fly ash bricks:** Burnt clay fly ash bricks replace a part of soil with fly ash for preparing the initial clay mix. This mix is then moulded and fired to produce burnt clay fly ash bricks.

3. **Hydrated fly ash bricks:** Hydrated bricks are also known as pulverized fuel ash-lime bricks, as lime, fly ash, and bottom ash are the primary constituents of the bricks. In practice the lime can also be substituted by cement. Hydrated fly ash bricks do not require firing and gain strength by pozzolanic reaction between lime and fly ash.

Independent of material used the brick should satisfy the specified values of compressive strength, water absorption, efflorescence and dimensional tolerance. Although the material is a major differentiator for bricks, the primary basis of classification is compressive strength. Bricks are designated in class as per the average compressive strength of the bricks. For all commercial purposes bricks are sold by their class designation and not material used. Table 1.9 shows the adopted classification for bricks based on their compressive strength.

Table 1.9: Classification of bricks

Class designation	Average compressive strength not less than (N/mm^2)
35	35.0
30	30.0
25	25.0
20	20.0
17.5	17.5
15	15.0
12.5	12.5
10	10.0
7.5	7.5
5	5.0
3.5	3.5

1.7 GLASS

Glass can be defined as an amorphous (non-crystalline) liquid, which has been super cooled to a solid like state. Glass is a transparent waterproof engineering material primarily used for aesthetics. The main application of glass in a building is to connect the outside and inside world visually. It is weather and corrosion-resistant, UV ray resistant, easily mouldable at the manufacturing stage, and also a good insulator of electricity. However, it is highly fragile having very low impact resistance, costly building material owing to its high energy-consuming manufacturing process and melts in high temperatures.

Glass is produced using the 'melt processing' process as shown in Fig. 1.8. In this process, raw materials sand, lime and soda/potash are mixed over 1000 °C to prepare a homogeneous form of molten glass. Other raw materials like lead, iron oxides, and borax are added to modify hardness and give colour to the glass. The molten glass is then shaped and cooled to produce the desired form of glass.

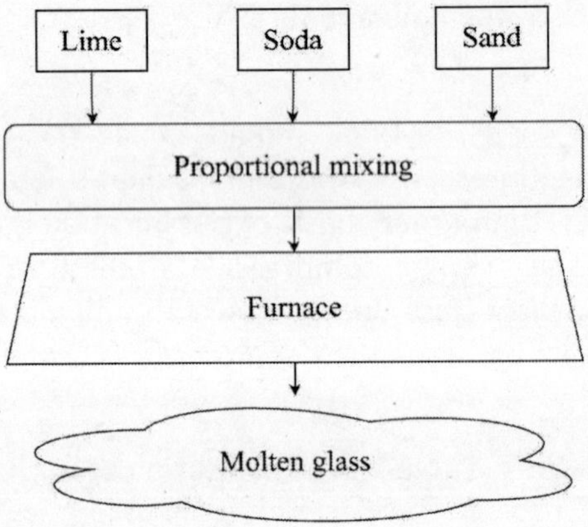

Fig. 1.8: Glass manufacturing process

Different types of glass:

Glass can be processed and engineered to exhibit different physical, chemical, and optical properties. The four most commercially used glasses are given below, also depicted in Fig. 1.9.

Soda-lime silicates: Soda-lime silicate glasses are used to make food and beverage containers, decorative items like window panes, showpieces etc. They account for majority of the commercial glass manufactured. A typical soda-lime glass contains 71 to 75 % silicate (SiO_2), 12 to 16 % soda (Na_2O), 10 to 15 % and small quantities of other substances such as dyes.

Borosilicate glass: Developed by a German glass-maker, it is one of the most durable and shock-resistant types of glass used. It can be used in environments of harsh temperatures and chemicals, like laboratories, jet fuel containers, explosion-proof lighting, etc. Borosilicate glass typically contains 70 to 80 % SiO_2, 7 to 13 % B_2O_3 and 4 to 8 % of Na_2O and K_2O.

Lead silicate glass: Lead silicate glass or lead glass is commonly known as flint glass. It is made of silica, lead and potassium. It is colourless, has a shining appearance and not affected by temperature. Hence, it can be used to make electric bulbs, optical glass, ornamental glass etc.

Phosphate glass: Its special property is to resist the effects of hydrofluoric acids, but on the other hand, has a low resistance to corrosion. It is well suited for doping with other colorants, hence, used to produce coloured glasses in many military, medical and scientific operations. In a typical phosphate glass major constituent is P_2O_5, in place of SiO_2.

Soda-lime silicates | Borosilicate glass | Lead silicate glass | Phosphate glass

Fig. 1.9: Examples of glasses

(Source: www.bvglas.de/en/about-glass/glass-is-a-multitalented-material/types-of-glass/; www.indiamart.com/proddetail/pyrex-glass-beaker-20611087212.html; www.pinterest.com/pin/552394710517467434/; www.facebook.com/smartelements/posts/iridium-doped-barium-phosphate-glassdoping-glasses-with-iridium-is-a-very-compli/2029530077122885/)

In addition to the type of glass based on its composition, glasses are also identified on the basis of their commercially used forms. Some of the commonly used forms of glasses are as follows:

Sheet glass: Used for glazing doors, windows, partitions and to make showcases.

Plate glass: Superior to sheet glass in a way that it has a parallel distortion-free surface.

Tempered glass: Plate glass when reheated and quenched increases the strength by 3 to 5 times and forms tempered glass. Tempered glass can partially resist bending stresses.

Wired glass: Embedding wire mesh into the mould before the casting of molten glass makes it suitable for fire-resisting doors and windows. The pieces of such glass don't get scattered during breakage.

Obscured glass: Surfaces of the glass are made opaque in nature by polishing, sand-blasting, etching of hydrofluoric acid, and making wavy and triangular patterns on the surface during casting.

Laminated glass: Also called safety glass, laminated glass is made by sandwiching layers of polyvinyl butyral in between layers of plate or sheet glass. It is commonly used as heat proof glass and bulletproof glass.

Coloured glass: Made by the addition of metal oxides to molten glass: opal glass, enamel glass, optical glass, blue glass, etc.

1.8 TIMBER

Any wood used for engineering purposes is termed as timber. Timber still makes a cost-effective construction material to use in remote locations, where other building materials are not readily available. Timber, as a construction material, offers good thermal insulation, better sound absorption, electrical resistance, shock load absorption and is easy to customize or work with.

A wood must possess certain characteristics to be classified as good timber, such as:

- Narrow annular rings and compact medullary rays.
- Dark-colour with uniformity in texture.
- Moisture content must be at most 12% during seasoning.

1.8.1 Classification of Timber

Timber can be classified in numerous ways with reference to IS codes:

1. On the basis of availability: Availability of timber is based on the information by the forest department and is classified as per IS 399, as shown in Table 1.10.

Table 1.10: Classification of timber based on availability

Class	**Availability per year**
Most common (X)	More than 1000 tonnes
Common (Y)	250 to 1000 tonnes
Less common (Z)	Less than 250 tonnes

2. On the basis of durability: Durability of timber is determined by testing the performance of timber in specific conditions, known as graveyard test, and is not the actual life expectancy of timber. On the basis of its average useful life under graveyard test, timber is classified as per IS 399, as shown in Table 1.11.

Table 1.11: Classification of timber based on durability

Durability	**Average useful life (in years)**
High	More than 10 years
Moderate	5 to 10 years
Low	Less than 5 years

3. On the basis of grading of sawed products: Presence of defects in timber decreases its strength. On the basis of reduction in basic strength, resulting from permissible defects, the sawed products of timber can be classified as per IS 1331, as shown in Table 1.12.

Table 1.12: Classification of timber based on strength reduction of sawed products

Grade	Estimated strength reduction after conversion (%)
Grade 1	Not more than 12.5%
Grade 2	Not more than 25%
Grade 3	Not more than 37.5%

4. On the basis of Modulus of Elasticity (E): According to IS 3629, timber can be grouped on the basis of modulus of elasticity, as shown in Table 1.13.

Table 1.13: Classification of timber based on Modulus of Elasticity

Class	E (kPa)
A	More than 12.5 kPa
B	9.8 to 12.5 kPa
C	Less than 9.8 kPa

5. On the basis of seasoning: Rate of drying during the seasoning often affects the quality of timber. Rapid drying in a refractory timber can result in cracking and splitting, and are difficult to season. Based on the rate or ease of seasoning IS 1141 classifies different timbers as listed in Table 1.14.

Table 1.14: Classification of timber based on seasoning ability

Class	Refractory property
A	Highly refractory
B	Moderately refractory
C	Non- refractory

6. On the basis of treatability: Classification on the basis of treatability, is defined by the resistance offered by the heartwood of the given species to preservative fluids under specified condition. Classification is as per IS 399, and can be listed as follows:

   - Heartwood easily treatable
   - Heartwood is treatable, but complete preservative penetration by applying pressure of 10.5 kg/cm^2 at 80 to 90 °C is not possible.
   - Heartwood is partially treatable
   - Heartwood refractory to treatment
   - Heartwood very refractory to treatment, no penetration is possible

1.8.2 Processing of timber

Processing of timber consists of the following steps in sequence:

Felling: It should be done at the time of maturity (when the age is around 50 to 100 years), done just above the root for maximum utilization, leading to economic use, and done when the moisture content is minimum.

Conversion: The process of making the timber customized for our utilization is called conversion. This conversion is done by the method of sawing.

Seasoning: Moisture should be removed up to a maximum limit of 12%. This seasoning results in low weight, thus reducing transportation costs. Also, seasoning results in a dimensionally stable cross-section, free from any cracks. It also makes the section denser, compact, and strong, hence improving its mechanical properties. Finally, seasoning makes the surface of timber suitable for finishing purposes.

Preservation: Timber is made of organic compounds, hence, liable to damage from factors like insects, dampness, fungi and fire. To ensure its appropriate utilization, timber must be preserved against such damage causing factors. For this, different coatings are applied on the surface of the timber for example creosote, coal tar, anti-pyrites, solignum paints, etc.

1.8.3 Defects of timber

Several factors during the life of tree or during the processing of timbers can result in defects in timber. Some of the commonly observed defects and their causes are listed as follows:

1. *Defects due to natural forces:* Following defects can be caused by natural natural forces during the life of tree

   Checks: Fine cracks indicating separation of fibres along the grain, not extending through the piece from one surface to another.

   Shakes: Cracks in timber due to excessive heat, wind or frost during the growth of the tree. Depending upon their shapes and positions, shakes are classified as cup shakes, heart shakes, ring shakes, star shake.

   Rind gall: Due to imperfect cutting of branches during the growth of tree, the tree may be wounded. To heal the wounds. curved swelling takes place. This defective portion is called rind gall

   Knots: Stumps of broken branches of the tree during the growth of the tree. They are categorized as live knot and dead knot. In the live knot, fibres are firmly held by the surrounding wood, whereas fibres are not held firmly in a dead knot.

   Twisted fibres: Caused by turning of the trunk of a tree when young in the direction of the wind.

2. *Defects due to conversion process:* During conversion defects can occur in the form of chip marks, diagonal grain, torn grains and wane.

3. *Defects due to insects:* Insects cause decay in timber. Beetles, marine borers and termites are some of the commonly found insects which can cause decay in timber.
4. *Defects due to fungi:* Following defects are caused by fungi

Druxiness: Also called white rot, it leaves thread-like white residual cellulose.

Foxiness: Reddish-brown stain over pith, due to poor ventilation.

Brown rot: The wood sugar gets decayed and the wood shrinks, showing brown coloration with cubical-shaped cracks.

Dry rot: Timber gets converted to dry powder when air circulation is not possible in damped places.

Wet rot: Timber becomes greyish in colour, due to alternate wetting and drying conditions.

Heart rot: Occurs when a branch come out of tree due to attack of atmospheric agents.

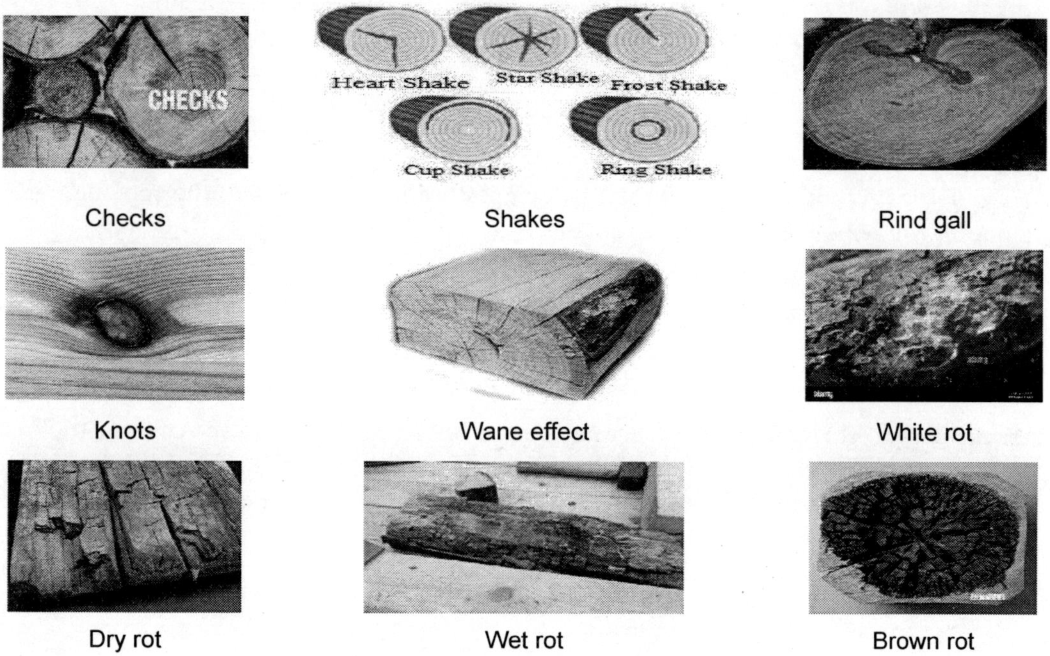

Fig. 1.10: Defects in timber

(Source: www.youtube.com/watch?v=9zT3qaZJxlw; www.kofastudy.com/courses/jss3-basic-technology-1st-term/lessons/materials-processing-timber-ii-week-2/topic/defect-in-timber/; www.buildersmart.in/blogs/defects-in-timber; www.kaltimber.com/blog/2018/3/5/how-to-identify-lumber-defects; www.eximcorp.co.in/wood-defects/wood-defects.pdf; www.alamy.com/stock-photo-timber-with-white-rot-49991927.html; www.safeguardeurope.com/applications/dry-rot; www.designingbuildings.co.uk/wiki/Wet_rot; www.researchgate.net/figure/Brown-rotted-wood-decay-from-the-collection-of-Holzforschung-Muenchen-Technical_fig3_346421380)

1.8.4 Common forms of timber

Common forms of timber available commercially, along with respective images as shown in Fig. 1.11, are as follows:

Veneer: Thinnest layers of timber cut out from a decorative material and used to make other timber forms.

Plywood: These are made by joining alternate thin layers of veneers in orthogonal directions together with adhesives. These are used to make partition walls, ceilings, panel walls, formwork for concrete, etc.

Fibre-boards: These are made from wood fibres boiled in hot water and natural gum removed from them using pressure. Then they are pressurized from top and bottom to make the fibre boards. These are generally used as insulating materials, and soundproof rooms.

Impreg timbers: Veneers are dipped in resin solutions like phenol-formaldehyde, then heated to dry them up. This timber has good resistance against moisture ingress, weathering, and electrical conductivity and also provides an aesthetic appearance, hence, used for decorative furniture and products.

Compreg timbers: Similar to impreg timbers, they are cured under pressure, hence have good quality control and are stronger than their impreg counterparts.

Hard boards: Wood pulp is compressed by pressure into boards and the top surface is made smooth, while the bottom surface remains rough.

Glulam: It means glued and laminated wood. Solid veneers are glued to form sheets and then laminated with suitable resins. These are generally used in chemical factories, long roofs, etc.

Chipboard: Made of powdered wood or rice husk ash which are dissolved in resins and pressure cured to make chipboards, also called particle boards.

Block Board: Leftover wood strips are glued together into a solid form and covered top to bottom by veneers, the thickness of the block doesn't exceed 25 mm. If the thickness is less than 7 mm, it is called Lamin board. These are also used for partitions, river crafts, railway carriages, etc.

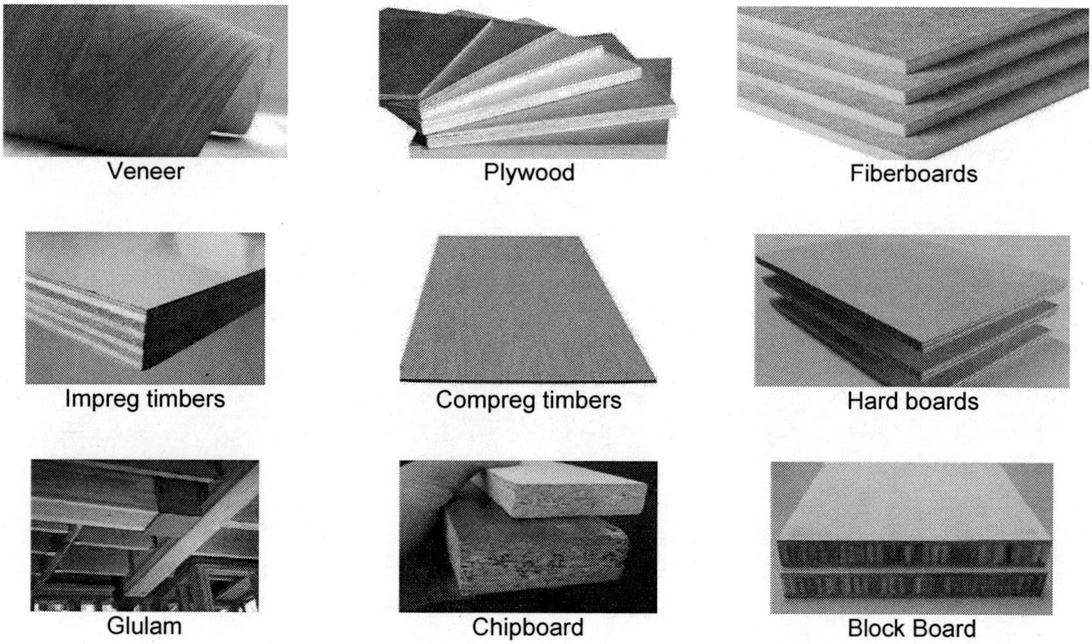

Fig. 1.11: Some commercially used timber

(Source: https://medium.com/@rohitgurjar009/different-types-of-industrial-timber-4b4345329250)

1.9 BITUMEN

Bitumen is a non-crystalline viscous binder material derived by fractional distillation of crude oil. Bitumen is substantially soluble in carbon disulphide (CS_2). Bitumen should not be confused with tar, which has a similar physical appearance as bitumen. Bitumen is different from tar, as the latter is obtained by destructive distillation of coal/wood and is soluble in toluene. Bitumen is classified as per its grade, which is defined by the viscosity value of bitumen at 60 °C.

Bitumen is generally obtained from heavy crude oils with high sulphur content, as depicted in Fig. 1.12. Bitumen refining starts by separating lighter fractions from the residues. This may be done by using atmospheric distillation followed by vacuum distillation. Atmospheric distillation separates lighter fractions from non-boiling matter (also called atmospheric residue) at the bottom. This residue is then used to separate lubricants from it without any change in chemical structure. This is done by using certain solvents, and this helps in the production of bitumen of different grades of penetration. This bitumen is then oxidized by blowing heated air to alter the physical properties for commercial use. Two types of bitumen are hence obtained - oxidized type, which is used as roofing materials, and air-rectified type, which is used for paving applications.

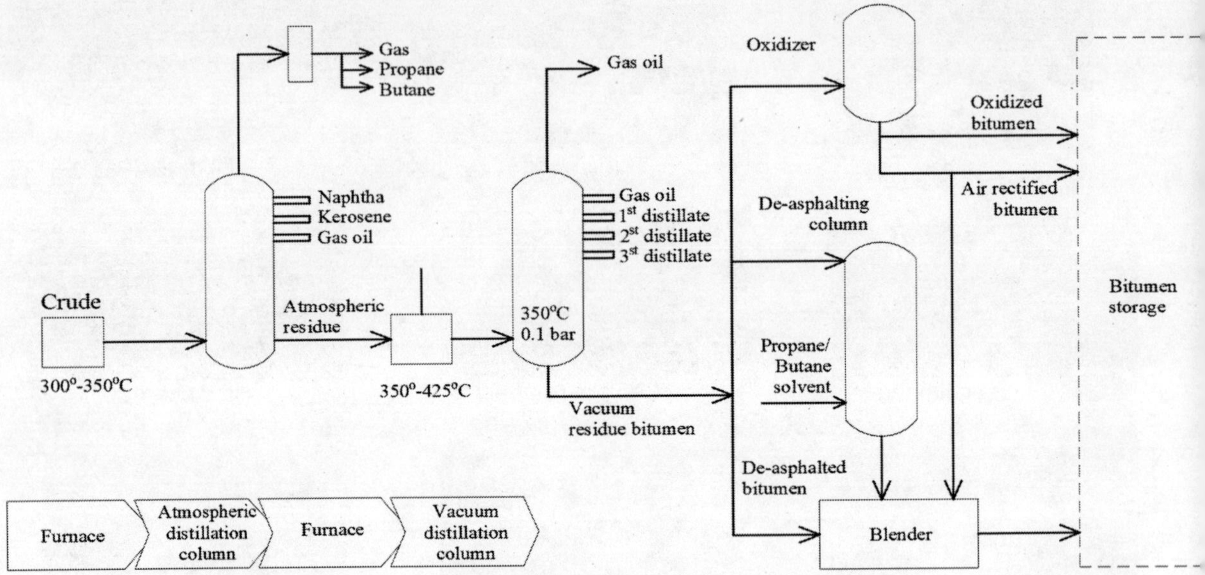

Fig. 1.12: Typical bitumen manufacturing process

Different types of bitumen:

The following are the different forms of bitumen available commercially, as shown in Fig. 1.13:

Penetration Grade Bitumen: Penetration grade bitumen is considered as the standard form of bitumen, produced for different viscosity values. This type of bitumen grade is based on penetration values. For example, 80/100 grade signifies 8 to 10 mm penetration at 25 °C, for a 5-second water bath.

Oxidized Bitumen Grades: Penetration-grade bitumen is further treated with processed air at a controlled temperature, making it soft. Thus, compounds of higher molecular weight are obtained which have low penetration grade and higher softening point.

Cut-back Bitumen: Viscosity of penetration grade bitumen is temporarily reduced (cut-back) by the addition of volatile diluents for easy workability during laying. Once the laying is done, the volatile matter evaporates, and the viscosity is regained back. Hence, this type of bitumen is best suited in colder regions due to the tendency of penetration bitumen to precipitate down frequently. If the viscosity is reduced by naphthalene, it is called rapid curing bitumen (naphthalene rapidly evaporates). If kerosene or low boiling point diluents are added, they are called medium and slow curing cutback bitumen respectively.

Bitumen Emulsion: It is a two-phase system made using two immiscible liquids. Bitumen, due to its sticky nature, tends to flocculate and settle down at the bottom of the water. To prevent this, long hydrocarbon chains with anionic or cationic endings are used as emulsifiers/ dispersing agents (generally soap solutions). Hence, due to charge repulsion, bitumen globules don't form flocs and do not settle down. Once the water evaporates due to heat, the bitumen hardens. If the evaporation of water is slow, it is called slow setting bitumen emulsion. Similarly, there are medium setting and rapid setting bitumen emulsions as well. Thus, this type of bitumen is best suited for repair and patch works in rainy seasons.

Polymer Modified Bitumen (PMB): Around 2 to 8% polymers are added to the bitumen to impart better viscoelastic properties, hence, best suited for roads that are to be designed for heavy-duty vehicles and extreme weather conditions. The addition of styrene-butadiene rubber (SBR) improves fatigue strength, compressive strength and waterproofing ability and lowers temperature sensitivity. The use of PMB can result in the requirement of a minimal pavement thickness, thus helping in cost-cutting.

Penetration Grade Bitumen — Oxidized Bitumen Grades — Polymer Modified Bitumen

Cut-back Bitumen — Bitumen Emulsion

Fig. 1.13: Commercially available bitumen forms

(Source: https://rahabitumen.com/penetration-grade-bitumen/; www.ariyancorp.com/oxidized-bitumen/; www.baumerk.com/en/blog/what-is-polymer-modified-bitumen-sheet-and-what-is-it-used-for; www.asiabitumen.com/product/cutback-bitumen-mc-250/; https://rahabitumen.com/bitumen-emulsion/)

1.10 ASPHALTIC MATERIALS

Asphalts, in general, can be defined as a composite mix prepared using aggregates, stone dust as inert filler material, and bitumen as a binder material. Asphalt can be categorized broadly in the following ways:

Natural Asphalt: Lake asphalt and rock asphalt are the two types of naturally available asphalt sources obtained from depths of 3 to 60 m beside fossilized lakes and bituminous rocks respectively. They are used as additives in asphalt and road paving, oil and gas drilling fluids, ink and paints manufacturing, and also in foundry sectors. It can make a road less susceptible to temperature variations and deformations, by making the cross-section of the road thinner. It is an environment-friendly additive compared to polymer additives, giving nearly the same properties like thermal resistance, waterproofing, and improved fatigue strength.

Mastic asphalt: Mastic asphalt or artificial asphalt is a form of composite material. It is prepared by mixing bitumen (as binder material), aggregates and filler materials like limestone powder. It behaves as solid or semi-solid in normal temperatures and sufficiently fluid when brought to high temperatures. Mastic asphalt is used in pavement constructions and water proofing.

Asphalt cement: It is a strong versatile weather and chemical-resistant binding material, mostly used to produce concrete for runways. It is a mixture of bitumen and asphalt with flux oils that have binding properties.

Cutback asphalt: It is produced by a mixture of asphalt cement with petroleum solvents, which are volatile in nature. After laying of the asphalt cement, the volatile components get evaporated and leave behind a solid residue of asphalt cement pavement.

Asphalt emulsion: It is produced by the mixture of asphalt cement with emulsifiers and water, which prevents flocculation of asphalt cement particles and hence prevents settling down in the water. Thus, like bitumen emulsion, these are also used in areas of heavy rains, for repair and patch-up works.

Asphalt materials have become much popular in the past few decades, owing to their versatility in use. For example, roads paved inside tunnels are made of asphalt cement concrete, due to their property of waterproofing, gaseous im-permeability, and ability to withstand heavy traffic and fatigue loads. Bridge decks are topped with asphalt surfaces, due to their thin cross-sections, which ultimately reduce the dead load on the decks. Asphalt cement concrete is used to produce sleepers for railway tracks, due to their ability to withstand dynamic loads. The use of such sleepers has been found useful for high-speed tracks since the wearing effect on the sleepers are less, and reduce vibrations and noise, hence, reducing the need for maintenance works. Runways, aprons, and taxiways are paved with asphalt surfaces due to their enhanced performance to withstand heavy wheel loads, and sufficient friction during landing operations; the roads become more waterproof, hence preventing chances of waterlogging.

1.11 STEEL

Among metallic materials, steel is the most commonly used material for construction of structures. Steel is the alloy of iron prepared using materials like carbon, chromium and nickel. One can obtain the desired modification in properties of steel like strength, ductility, and hardness by varying different factors like carbon content, number of alloying materials, heat treatments and mechanical works.

1.11.1 Manufacturing of steel

The prominent steel-making processes are:

1. Cementation process
2. Crucible process
3. Bessemer process
4. Open hearth process
5. Linz and Donawitz (L.D.) process
6. Kaldo process
7. Duplex process
8. Electric process

The earliest form of steel was made using a steel making process called cementation in the 17th century. In this process, the bars of wrought iron are layered with powdered charcoal and heated at high temperatures for about 5 to 15 days. The steel produced by this process is named *"blister steel"*, due to the presence of fissures and cavities in the structure Later, crucible process became popular, which produced steel by heating the fragments of blister steel or wrought iron bars in a fire clay crucible with charcoal. In 1856. The Bessemer process was invented and is the basis for most steel-making methods used today. In this process, hot air bursts are used to both purify the iron and agitate the alloy to ensure proper mixing. Open hearth method or the Siemen's-Martin technique, uses coal gas (for carbon) along with hot air to produce more homogeneous steel than Bessemer process. The L.D. process is a variation of the Bessemer process, where a jet of pure oxygen is used for purifying the iron at high temperatures. Kaldo process is a variation of L.D. process and uses a rotating Kaldo-LD converter for feeding oxygen. The Duplex process combines two different processes to produce steel, like the Bessemer process and Open-hearth process. Typically, in duplex process, the first process is used to partially refine the metal and lower the melting costs, while the second process is used for good quality refining. The electric process uses electric arc to melt iron for steel making. Despite higher cost, the electric process is used for lack of smoke emission and good quality of resulting steel. Most of the modern steel making processes are either a modification on the Bessemer process or electric process.

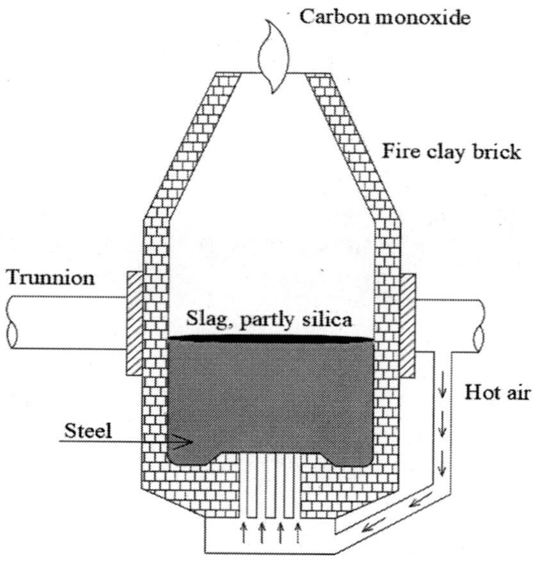

Fig. 1.14: Bessemer converter for the manufacturing of steel

1.11.2 Properties of steel

On the basis of properties of steel, there are numerous advantages of using steel as a construction material. The preferred properties of steel include tensile strength, ductility, durability, compressive strength, hardness, toughness, yield strength, fatigue strength and creep. Among these properties, behaviour under tension is considered more important for steel as an engineering material. Fig. 1.15 depicts a typical stress-strain curve for mild steel. The values of numerous quantities are directly provided in this figure.

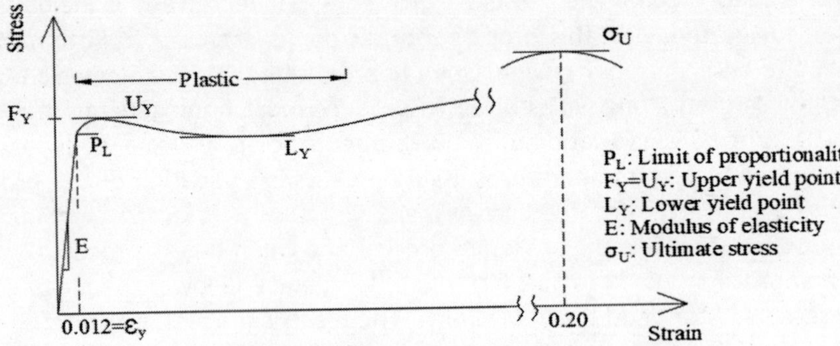

Fig. 1.15: Typical stress-strain curve for mild steel

At low strain values the stress and strain are linearly proportional, up to the limit of proportionality. The Young's Modulus, also known as slope of stress strain diagram, up to limit of proportionality is defined as:

$$E = \sigma / \epsilon \tag{1.4}$$

At low strain values the steel behaves as an elastic material and reverts to its original state when loading is removed. On further increase in strain, steel starts to deform permanently, even after the load is removed. The stress measured at the point during the tensile test when the plastic deformation, i.e., permanent deformation, begins is known as upper yield point. On further application of strain, the lowest stress value for which plastic deformation escalates is known as lower yield point.

At higher strain values, the stress again increases with the increase in strain, until the specimen attains a maximum value, called the ultimate stress. In this region steel undergoes strain hardening and offers further resistance to applied loads. For most engineering applications modulus of elasticity, yield stress and ultimate tensile stress are the important parameters of steel.

Factors affecting the properties of steel

The factors influencing the properties of steel include chemical composition, principal impurities, heat treatment, and mechanical working.

Chemical Compositions:

Carbon has a major role on the properties of steel. Carbon provides hardness, and strength to steel. Elastic limit, yield strength and ultimate strength of steel increase with carbon content. Its compressive and shear strength also rises along with its carbon content. However, excess carbon may lead to increased brittleness and rusting. Increase in carbon content also reduces the magnetic permeability, malleability and ductility of steel.

Effects of Principal Impurities on Steel:

While making steel or iron, impurities cannot be fully eliminated. The final product always contains traces of the metallic impurities, like silicon, manganese, sulphur, and phosphorus; in addition to primary constituents, i.e., iron and carbon. Copper and arsenic are also present in uneven trace levels. In well manufactured steel, these impurities generally range from 0.2 to 1.0%, and their influence on the composition of steel is negligible.

- One common impurity that cannot be eliminated during the production process is phosphorus. Phosphorus can have both beneficial as well as harmful effects on steel. It increases the steel's yield point, ultimate strength, and hardness. Additionally, 0.1% phosphorus lowers shock resistance, making the metal cold short (i.e., brittle when cold). On the other hand, it encourages grain expansion leading to brittleness. The inclusion of 0.3-0.5% phosphorus reduces the ductility of low-carbon steel.
- Other common impurities are of silicon and manganese; which are mostly introduced to improve the manufacturing of steel. Silicon is frequently added to molten metal to reduce blow holes and eliminate oxygen. Silicon is generally present in small concentrations and tend to improve its strength.
- Manganese is added to clean the molten metal by pulling most of the unwanted impurities into the slag, due to its strong affinity for oxygen and sulphur. The effects of sulphur are lessened by manganese, which also promotes hardness and tensile strength. In case of a high amount of manganese, the extra manganese produces carbide and serves as a hardener.
- Iron sulphide (FeS) is an impurity which creates brittle networks in steel. on heating, iron sulphide reduces cohesion amongst nearby grains and make steel difficult to forge. Manganese is added to remove sulphur and resolve this impurity.
- Copper makes steel corrosion resistant when present in a small amount.
- Arsenic induces strength and brittleness.

Heat Treatment:

Heat treatment is used to modify the properties of steel. In heat treatment steel is exposed to predefined cycles of heating and cooling. A steel of given composition may be made soft and ductile by one heat treatment, and the same steel may be made relatively hard and strong by another. Heat treatment affects the metallographic properties, solubility of ingredients in relation to one another, crystallization in terms of shape or degree of aggregation, and adds or removes internal tensions in the metal. It can also be used for removal of gaseous impurities and refining the grain.

There are various methods of heat treatment as described below:

Hardening: In this method, steel is heated and held above a critical temperature, known as transformation range, until the state of equilibrium is achieved. After this the steel is rapidly cooled (quenching). With this process, hardness can be increased. Hardened steel is brittle and cannot be used for many engineering applications.

Annealing: It is a process of heating and gradual cooling, used to make the steel soft. The procedure involves heating the steel to a specific temperature, defined by its transformation range, and then gradually cooling. The precise heating temperature varies depending on the steel's composition and the nature of annealing to be performed. Annealing induces softness, malleability, and ductility. It can also be used for relieving stresses formed during cold working.

Normalizing: In this method steel is heated above the transformation range and then air cooled. While the exact effect of this method depends on the rate of cooling, the main purpose of this method is to remove the effects of any previous heat treatments.

Tempering: In this method, hardened or normalized steel is reheated to a high temperature, below a certain critical temperature, and then cooled. This process is used to enhance the toughness and reduce the hardness of material. Through tempering, the ductility increases and brittleness of steel decreases.

Mechanical Working:

Mechanical working is used to shape the metal by means of external forces. Stress higher than yield stress and less than ultimate tensile stress is applied to obtain the desired shape through plastic deformations. Mechanical working introduces permanent deformations, which increases the internal stress of steel.

1.11.3 Classification of Steel

Based on carbon content:

Among various factors, carbon content is one of the most commonly used parameters for classifying different types of steel, as shown in Table 1.15.

Table 1.15: Typical carbon content for different steel types

Steel type	Carbon content (%)
Mild steel (low carbon steel)	< 0.30
Medium carbon steel	0.35-0.60
High carbon steel	0.65-1.05
High carbon low alloy steel	1.10-1.20

Mild steel: Mild steel is also called low carbon steel. It is stronger and more elastic than wrought iron, malleable and ductile. It is forgeable and weldable but challenging to temper and harden. A typical mild steel can have a specific gravity of about 7.80, with ultimate compressive and tensile strengths in the range of 800-1200 N/mm^2 and 600-800 N/mm^2, respectively. Rolled sections, reinforcing bars, roof coverings, sheet piles, and railroad tracks are all made of mild steel.

Medium carbon steel: Medium carbon steel has 0.35 to 0.6% carbon content. The ultimate tensile strength is typically in the range of 450 - 2730 N/mm^2. It is used as prestressing wire in construction of prestressed concrete.

High carbon steel: High carbon steel has a carbon content that ranges from 0.65 to 1.05%. It is also called hard steel. Compared to mild steel, it is more elastic and robust. It is challenging to forge and weld. High carbon steel can have ultimate compressive strength and ultimate tensile strength in the range of 1320-3100 N/mm^2 and 161-3200 N/mm^2, respectively. Specific gravity of high carbon steel is about 7.90. Cement concrete and prestressed concrete members are usually reinforced with high carbon steel. It is also utilized to make tools and machine components because it can withstand shocks and vibrations.

Based on geometrical shape (Rolled Steel Sections):

Steel can be rolled into a variety of shapes and sizes in rolling mills, to meet the desired application requirements. Rolled steel sections are commonly identified using cross-sectional shapes. Some commonly used rolled steel sections are as follows:

Angle sections: An angle-section is designated by its leg lengths and thickness. For example, Indian standard angle section ISA 40 × 25 × 6 mm means, the section is an unequal angle with legs 40 mm and 25 mm in length and thickness of the legs 6 mm.

I-section: It is designated by overall depth and weight, e.g., ISLB 500 @ 735.7 N/m means, the I-section is 500 mm deep, and the self-weight is 735.7 N per metre length.

Channel section: It is designated by overall depth and weight. For instance, ISLC 350 @ 380.6 N/m means that the channel section is 350 mm deep, and the self-weight is 380.6 N per meter length.

T-section: It is designated by overall depth and weight, e.g., ISNT 125 @ 268.8 N/m, means the T- section is 125 mm deep and the self-weight is 268.8 N per meter length.

Steel flats: They are designated by width and thickness of the section, e.g., 30 ISF 10 mm means the flat is 30 mm wide and 10 mm thick.

Steel plates: They are designated by length, width, and thickness, e.g., ISPL 2000 mm × 1000 mm × 8 mm, means the plate is 2000 mm long, 1000 mm wide and 8 mm thick.

Steel sheets and strips: They are designated by length, width, and thickness, e.g., ISSH 2000 mm × 600 mm × 4 mm, which means the sheet is 2000 mm long, 600 mm wide, and 4 mm thick.

Bars: They are designated by the shape and defining dimension, e.g., ISRO 5 means a round bar of 5 mm diameter, and ISSQ 10 means a square bar of side 10 mm.

1.11.4 Reinforcing Steel Bars

The reinforcing steel bars are most commonly used as reinforcement in cement concrete structures. In general application, high tensile strength is expected from the reinforcing bars. Use of high carbon steel, mechanical strengthening, or heat treatment are all possible ways to improve the tensile strength of steel bars. Greater yield strength of the reinforcing steel bars reduces the need for steel, which decreases the cost of the reinforcement and its fitting. These bars are specified on the basis of their diameter, tensile strength, and chemical composition. On the basis of their shape, these bars can be plain or deformed.

Classification:

Steel for reinforcing bars can be classified according to their tensile strength, also designated as their grade. The different types and grades of reinforcing bars are given in Table 1.16.

Table 1.16: Types and Grades of Reinforcing Bars

Types of steel	Bar diameter	Yield stress or 0.2% proof stress	Minimum elongation
1. Mild steel grade-I	Up to and including 20 mm	250 N/mm^2	23%
	Over 20 mm up to and including 50 mm	240 N/mm^2	23%
2. Mild steel grade-II	Up to and including 20 mm	225 N/mm^2	23%
	Over 20 mm up to and including 50 mm	215 N/mm^2	23%
3. Medium tensile steel	Up to and including 16 mm	350 N/mm^2	20%
	Over 16 mm up to and including 32 mm	340 N/mm^2	20%
	Up to and including 50 mm	330 N/mm^2	20%
4. High yield strength deformed (HYSD) steel bars	All sizes	415 N/mm^2	14.5%
		500 N/mm^2	8%
		550 N/mm^2	6%
5. TMT Bars	All sizes	415 N/mm^2	22%
		500 N/mm^2	20%
		550 N/mm^2	18%

The earlier form of steel used for construction purposes was plain mild steel bars, designated as Fe 250. Despite their lower tensile strength, plain mild steel bars are still used for construction, as they do not show brittle failure. Also, better ductility of mild steel means, the bars are good for impacts and suddenly applied loads. Mild steel grade-II and medium tensile steel bars are used for higher tensile strength requirement.

HYSD bars are another type of widely used bars. HYSD bars include lugs, ribs, or other surface deformations that prevent the bar from moving or slipping along the longitudinal direction in concrete (Fig. 1.16). The surface's deformation promotes a stronger connection between the reinforcement and the concrete. There is no established yield point for these bars. HYSD bars significantly enhance yield, tensile, and bond strength when they are twisted, either by hot or cold working. In India, cold twisted deformed (CTD) bars are frequently used for construction projects, as HYSD bars.

Fig. 1.16: Typical HYSD (High yield strength deformed) bars

(Source: https://civilsnapshot.com/high-yield-strength-deformed-bars/)

Now days, for most engineering applications, mild steel bars and CTD bars have been replaced with thermally mechanically treated (TMT) bars. TMT bars have high tensile strength, similar as CTD bars, and offer good ductility, similar as mild steel bars. They have a definite yield point and can resist temperature up to 500 °C with no loss of strength. They have excellent bending ability due to its unique feature of uniform elongation. These bars can also be welded easily with other TMT bars or can be welded with CTD bars and don't induce strength loss in welding joints. Due to the high strength of the surface layer, these bars have an exceptionally high fatigue resistance under dynamic loading. Due to their thermal stability and complete lack of a cold-worked structural zone, thermally hardened reinforcing steel bars are better suited for usage in locations where there is a risk of fire. TMT bars are very formable because of their high percentage of consistent elongation property. They are highly used for high rise building and can be used in general buildings and bridges.

Thermo-mechanically treated high-strength corrosion resistant (TMT-HCR) rebars are one of the most recent innovations in steel reinforcing bars. The TMT-HCR rebars as the name suggests are corrosion resistant bars, which have exceptional resilience to harsh weather conditions. Even at temperatures as high as 600°C, TMT-HCR reinforcement bars demonstrate greater thermal resistance. TMT-HCR rebars have a number of benefits over conventional CTD rebars, including better high temperature thermal resistance, easier welding on site due to better ductility and bendability, increased strength, superior atmospheric and marine corrosion resistance, good weldability and no loss of strength at welded joints, and cost savings from using less steel. TMT-HCR rebars are widely used in a variety of fields, including those that include corrosion-prone coastal and marine environments, bridges, flyovers, dams, commercial and industrial buildings, and subterranean platforms.

1.11.5 Rusting and Corrosion

When steel is exposed to the atmosphere, the iron within the steel combines with the moisture and air to produce iron oxides, through electrochemical reactions. The resulting iron oxides dissolve into the surrounding solutions and steel loses its strength and other useful properties. This phenomenon is called corrosion. The process of corrosion is gradual, and can take many days before showing any visible surface damage. In this phenomenon, the volume of steel increases and rusts start peeling off from the swelled surface of steel. This corrosion is the major reason for the low life span of both concrete and steel structures. The chemical reactions involved in rusting are shown below.

$$Fe + O + 2CO_2 + H_2O \rightarrow Fe(HCO_3)_2 \tag{1.5}$$

$$2Fe(HCO_3)_2 + H_2O + O \rightarrow 2Fe(OH)CO_3 + 2CO_2 + 2H_2O\ Fe(OH)CO_3 + H_2O \rightarrow Fe(OH)_3 + CO_2 \tag{1.6}$$

There are several methods with which rusting can be prevented of which two of the most used methods are:

Enamelling: In this method, a flux is melted on the surface of steel in the furnace and then a second layer coating is done with more fusible glaze.

Galvanization: In this method iron is coated with a thin film of zinc, or similar sacrificial metal. The electrochemical reactions occur over zinc and prevent the steel from corrosion.

In addition to this several paints and protective coatings are also used to prevent or delay the corrosion. Paints and protective coating are applied on the exposed surfaces after construction.

1.11.6 Alloy Steel

In general, a single metal cannot exhibit all the qualities required in a metal to be employed as a construction material. Alloys are made by combining metals or metallic compounds to create new materials with desired qualities. In general sense, steel is made an alloy for making steel stronger with reduced treatment, and grant a specific characteristics like machinability, corrosion resistance, wear resistance, etc. Table 1.17 lists some of the most popular alloys of steel along with their characteristics and applications.

Table 1.17: Properties and Uses of Alloy Steels

Alloy steel	Typical Composition	Properties	Uses
1. Stainless steel	Chromium 10-30%	Very hard and tough, high elasticity and ultimate strength, acid and rust resistant, aesthetically pleasing	Ball bearings, dyes, crushing machines, medical use, razors
2. Nickel steel	Nickel more than 3.5%	More elastic, higher tensile strength, less brittle than mild steel, improved hardness and ductility	Automotive and airplane parts
3. Invar steel	Nickel 30–40%	Low coefficient of thermal expansion	Delicate instruments
4. Vanadium steel	Vanadium 0.1–2%	High tensile and yield strength; Resistance to softening at high temperatures	High speed tools, locomotive castings, auto parts, chassis
5. Tungsten steel	Tungsten 14–20%	High cutting hardness; Resistant to abrasion	Drilling machines, high speed tools
6. Manganese steel	Manganese 12–15%	Hard, tough, and strong; Difficult to machine; High electrical resistance	Points and crossing in railways, rollers, jaws of crushers, Heavy earth, and mining equipment
7. Molybdenum steel	Molybdenum 0.2–0.3%	Maintains tensile strength at high temperatures	Gears, axles, shafts

1.12 PAINTS AND VARNISHES

In simplified terminology, a liquid used for covering any surface is known as paint. Paint generates a thin coating (30-300 µm) on the surface which it is applied, after drying. Paints can be of different types, such as oil paints, enamel paints, cement paints, bituminous paints, and special paints like illuminating paints, chlorinated rubber paints (used to shield things from acid vapours), etc.

The purpose of the paint is to defend the coated surface from potential mechanical, chemical, and physical stressors as well as environmental and physical deterioration. Paints are also used for decorative purposes, by applying a smooth and vibrant finish. Other advantages of paints include check for water infiltration through RCC, check for the growth of germs and fungus that are unsanitary and give the walls an unsightly appearance, check for metal structure corrosion; and polish the surface for better aesthetic appeal.

1.12.1 Composition of oil paint

Base:

The main component of the paint is the base, which is typically a metallic oxide. It has binding qualities and lessens the shrinkage cracks formed on drying of paint film. Base makes the paint film opaque. White lead, red lead, zinc white, and titanium oxide are a few examples of base. Lead-based paints have better performance than other bases, but also have a negative impact on the environment. Hence, the use of such paints is not advised for use as top coats. Zinc white is resistant to corrosion and sulphate attack; hence it is preferred for use as top coat. While, titanium oxide is preferred as under coat due to its refractive index.

White lead: This is a carbonate of lead and forms the base of lead paints. It is highly used on wood and building surfaces. White lead is not used for metal surfaces, as it does not prevent corrosion.

Red lead: This is an oxide of lead and forms the base of lead paints. It is most suitable for painting iron surfaces and for providing a priming coat to wood surfaces. It solidifies in a short time with linseed oil and hence, it is used as a drier also.

Zinc white: This is an oxide of zinc and forms the base of all zinc paints. It is smooth, transparent, and non-poisonous. It is not discoloured when exposed to sulphur vapours. It is less durable and is difficult to work.

Oxide of iron: This is an oxide of iron and forms the base of all iron paints. The tint of paint varies from yellowish brown to black. It mixes easily with the vehicle. It is effective in preventing rusting of iron surfaces and is cheap and durable. It is generally used for priming coats of iron surfaces.

Titanium white: This material possesses intense opacity. It is non-poisonous and provides a thin transparent film. It is used for receiving the coat of an enamel.

Aluminium: This forms the bulk of aluminium paints. It keeps moisture content of wood surfaces practically the same and prevents cracking and warping of wood. It is generally used as a priming coat to new wood work.

Lithophone: This is a mixture of zinc sulphide and barytes. It is similar in appearance to oxide of zinc. It is cheap and can easily be applied on the surface. However, when exposed to daylight, it changes colour, hence used for interior works only.

Vehicle:

Vehicle, also referred to as a *binder*, is an oil that is combined with the base. It keeps the paint's ingredients in suspension, aids in spreading them throughout the area that needs to be painted, gives the paint film robustness, durability, water resistance as well as weather resistance and gloss to the painted surface, and serves as the paint's body. Examples include naturally occurring drying oils, including linseed, nut, poppy, and tung oils, animal, paint, artificial, and synthetic glues in glue paints, and air-slaking lime and polymer in lime water colours and polymer paints, respectively.

Pigments:

Pigments provide desired colour to the paint and help in covering up the flaws on the surface. They also shield the paint coating by deflecting the harmful ultraviolet radiation, which can cause oxidation of the paint film. They also help in reducing the permeability of the paint coat to strengthen its resistance against weathering. They have a wide range of characteristics, including covering strength, colour-holding ability, fineness, fire resistance, chemical stability, and weather resistance. The particles of pigments are of fine size, typically in the range of 0.1 to 5.0 microns. The paint film is strengthened by the fine size of pigment particles.

A pigment may be of natural or artificial origin. Examples of natural pigments include, natural white chalk, mastics, grey graphite, dried yellow ochre (a clay with over 15% iron oxide), etc. Examples of artificial pigments include titanium dioxide, zinc white, lead white ($2\ PbCO_3.Pb\ (OH)_2$), lithophone ($BaSO_4 + ZnS$), red lead, gas black soot, etc.

Solvents:

The oils used to thin and distribute paint are called solvents or thinners. Solvents make the paint more flowable and support its application. Once the paint has been applied the solvent evaporates and leave the paint coating on the surface. Petroleum spirit, naphtha, and turpentine oil are some examples of solvents. Turpentine, is most commonly used solvent, due to its potent solvent properties, outstanding flattening abilities, and perfect evaporation rate.

Driers:

Driers, commonly referred to as plasticizers, are substances that are added in a paint to speeding up the drying process of the vehicle. The amount of drier is generally kept below 8%, because too much will cause paint to lose its flexibility and flake. Letharge (oxidised lead, PbO), lead acetate, red lead (Pb_3O_4), manganese dioxide, cobalt, zinc, and lead chromate are a few examples of driers.

Adulterants:

Adulterants decrease the overall cost and weight of the paint. They also help in improving the durability. Adulterants also occasionally assist in keeping the pigment in suspension and lessen dry paint cracking. Examples of adulterants include silica, calcium carbonate, magnesium silicate, and barium sulphate. Silica is preferred in undercoats, as its roughness helps in the development of the bond with the subsequent coat.

1.12.2 Preparation of paint

The base is ground to a paste-like consistency in a tool called a muller. The base is stirred with a wooden spoon as small amounts of vehicle oil are intermittently added. For coloured paints, pigment and vehicle oil are separately blended using the same process as for base. Driers are also ground in vehicle oil. The three pastes are then mixed together. In this mixture, more vehicle oil is and continuously stirred, until the mixture softens and reaches the consistency of cream. After that, the mixture is run through a sieve or thin canvas to produce the final product, i.e., paint. Commercial paint is typically produced in multi-storey buildings and use gravity to move materials. Fig. 1.17 depicts a flow diagram of the production process for paint.

The quality of the ingredients, pigment volume concentration, careful mixing, grinding and straining are some of the factors that affect the quality of final paint.

Pigment volume concentration (PVC):

It is the volume concentration of the pigments represented as a percentage of the overall volume of the paint's non-volatile components.

The amount of a certain colour that may be added to the paint depends on the PVC value. To form a protective coating, the pigment must be sufficiently moist. The pigment particles must be thoroughly covered or wetted, which requires a suitable amount of binder. The binder should also cover the spaces between the pigment particles. The term *"critical pigment volume concentration"* (CPVC) refers to the point where there is just enough binder to moisten the pigment particles. There is enough pigment to be covered by binder below the CPVC, but not above it. Numerous physical and visual characteristics of paint alter at CPVC.

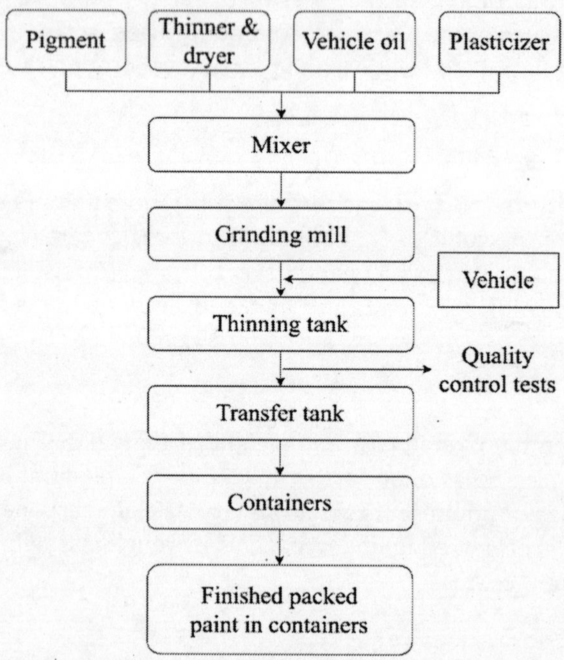

Fig. 1.17: Flow diagram of paint manufacture

Importance of pigment volume concentration: PVC affects several properties like gloss, washability, adherence, undertone, concealing power, permeability, and durability. As PVC content rises, quantity of binder decreases and air occupies the gaps. Above CPVC gloss decreases till paint turns flat, blistering declines and permeability increases. Table 1.18 shows the typical PVC value for different finish qualities. Lack of binder causes the pigment to become defective and the characteristics of paint to deteriorate. The paint film generated loses cohesiveness, durability and washability. Tensile strength, blistering, gloss, corrosion, permeability, enamel holdout and scrub resistance are some of the factors that are usually tested both above and below CPVC. Extenders also contribute to the amount of PVC in a paint. So, if extenders are being used, they should be used for substituting pigments and not in addition to them.

Table 1.18: Effect of change in PVC on expected finish quality of paints

Finish quality of paints	PVC
Flat	40 - 80 %
Egg shell	35 - 45 %
Satin	30 - 45 %
Semi-gloss	15 - 25 %
Gloss	0 - 15 %

1.12.3 Varnish

Varnish is generally defined as a uniform blend of resin in solvents like, turpentine, alcohol, or oil. In varnish, the solvents dry with time, leaving behind a firm, transparent resin layer that covers the surface. Varnish can be prepared using different types of resins and solvents. Table 1.19 lists some commonly used resins along with suitable solvent used for making varnish. Driers like letharge and lead acetate are also used increasing the speed of drying.

Table 1.19: Common resins and corresponding solvents used for making varnish

Resin	Solvents
Amber, copel, gum anime	Boiled linseed oil
Common resin, gum dammer, mastic	Turpentine
Lac, shellac, sand arch	Methylated spirit
Raw copal, cheaper types of resins	Wood naphtha

Varnish provides a protective coating and gloss to the surface of the wood. It brightens the appearance of the grain in wood and renders a brilliant appearance to the painted surface.

Ideally, the varnish should dry rapidly and present a uniform finished surface. The colour of varnish should not fade away when the surface is exposed to atmospheric actions. furthermore, the protecting film developed by varnish should be tough, hard, durable and free from shrink or cracks after drying.

Varnishing:

Varnishing is the process of applying varnish. It can be described as follows:

- Prepare the surface by making it smooth (by rubbing it with sandpaper) followed by cleaning.
- Cover the knots in wood. For these two different approaches, size knotting and patent knotting, are commonly used.
- Apply weak glue on the wood surface, then rub and clean the surface.
- Apply two coats of varnish.

Types of varnishes:

Varnishes can be classified as follows:

Oil Varnish: Linseed oil and resin, like copal and amber, are combined to create this varnish. Turpentine oil can be added in small amounts if the varnish proves to be unworkable. It is appropriate for both interior and exterior projects.

Spar Varnish: It receives its name from the fact that it is used on spars and other parts of a ship. It is not used indoors, as it has a sticky effect in warm weather.

Flat Varnish: It is created when elements like wax, metallic soap, or finely divided silica are added to varnish, which causes the surface to have a dull appearance.

Spirit Varnish: It is produced by dissolving soft resin, like lac or shellac, in a spirit. It dries out quite fast but lacks durability, hence it is more suitable for indoor applications. French polish, lacquer, and shellac varnish are some examples of spirit varnish.

Asphalt Varnish: It is produced by combining linseed oil with melted hard asphalt and a suitable thinner. It is generally applied over fabricated steel.

Water Varnish: It is made by dissolving shellac in hot water and then adding sufficient amounts of ammonia, borax, soda, or potash. It is employed in the varnishing of documents, maps and images.

1.13 ASBESTOS

Asbestos are naturally occurring fibrous minerals. The fibrous structure of asbestos can be broken down into fine fibres by application of mechanical force. Natural asbestos is of two types, acid resistant and non-acid resistant. Acid resistant asbestos include crocidolite asbestos, anthophyllite asbestos, amosite asbestos, actinolite asbestos, and tremolite asbestos. While, chrysotile asbestos is the only type of non-acid-resistant asbestos.

The reason for naturally occurring fibrous form of asbestos, is the strong lateral link between asbestos in one direction and weak in other directions. This also explains the tensile strength along the fibres, as well as fibre splitting under mechanical force. The asbestos fibres can be as fine as 10 to 30 nm in diameter, and are often fluffed together to form larger diameters. Asbestos fibres typically have a high tensile strength of 3000 N/mm^2. But when asbestos is fluffed, fibres are subjected to compression, impact, and mechanical actions which reduce the strength to an order of 600 to 800 N/mm^2, which is still higher than commercial HYSD bars.

Asbestos has a low thermal and electrical conductivity, a good alkali resistance, and a low acid resistance. Asbestos is highly resistant to fire, as it melts roughly at a temperature of 1550 °C and retains most of its strength up to 400 °C. Asbestos is commonly used in asbestos concrete sheets for roofing, false ceilings, panelling, partitions, wall linings, door panels, window panes, sign boards, closets, etc. Despite the numerous advantages, asbestos present a critical health risk. Asbestos fibres, when not bonded by concrete, have the tendency of becoming airborne and causing respiratory issues. This presents a challenge in both handling of asbestos during casting and repair of asbestos concrete. Recognizing the negative impacts, the use of asbestos has decreased significantly in the construction industry.

UNIT SUMMARY

Engineering materials and their classification

- Ceramic, Metal, Polymer, Composite and Others.

Different engineering materials

- Cement, M sand, Glass, Timber, Bitumen, Asphalt, Steel, Aluminium, Copper, Tin, Nickel, Paints and Varnish.
- Different variations of each engineering material.

Process of developing engineering materials

- Raw materials used and their properties.
- Different production techniques.
- Possible modifications and treatments with their effect on the final engineering material.

Properties of engineering materials

- Strength based – compressive, tensile, flexural, shear, split tensile, fatigue, impact, ductility
- Others - weight, density, colour, opacity, flowability, setting, hardening, thermal conductivity, acoustic, durability
- Important properties for different engineering materials

Application of engineering materials

EXERCISES

Multiple Choice Questions

1.1 For the manufacture of Portland cement, the proportions of raw material used, are

(a) Lime 17-25%, Silica 62-67%, Other ingredients 15%

(b) Lime 62-67%, Silica 17-25%, Other ingredients 15%

(c) Lime 40-45%, Silica 40-45%, Other ingredients 10-20%

(d) Lime 30-35%, Silica 30-35%, Other ingredients 30-40%

1.2 The initial setting time of cement is caused due to

(a) Tri-calcium silicate (b) Di-calcium silicate

(c) Tri-calcium aluminate (d) Tetra calcium alumino ferrite

1.3 Which of the following acts as flux in dry process manufacturing of cement

(a) Magnesium oxide (b) Alumina oxide (c) Ferric oxide (d) None of the above

1.4 What is the mix proportion (Cement: Fine aggregate: Coarse aggregate) for M15 grade concrete

(a) 1:4:8 (b) 1:2:4 (c) 1:3:6 (d) 1:1.5:3

1.5 The unit weight of plain concrete is generally taken as

(a) 24 kN/m^3 (b) 25 kN/m^3 (c) 20 kN/m^3 (d) 30 kN/m^3

1.6 As per IS 456 (2000), the mandatory minimum concrete grade required for RCC is

(a) M10 (b) M15 (c) M20 (d) M25

1.7 Glass is defined as which form of ceramic mixture

(a) Crystalline (b) Amorphous (c) Both (a) and (b) (d)None of the above

1.8 Classification of class A timber based on modulus of Elasticity (kPa)

(a) Less than 8.5kPa (b) Less than 9.8kPa

(c) 9.8 kPa- 12.5kPa (d) More than 12.5kPa

1.9 The thinnest layer of timber cutout is known as

(a) Fibre-boards (b) Impreg timbers (c) Vineer (d) Compreg timbers

1.10 The grade of bitumen is designated by its viscosity at

(a) 40°C (b) 45°C (c) 60°C (d) 80°C

1.11 Bitumen may be dissolved in

(a) Water (b) Carbon disulphide (c) Carbon dioxide (d) Sodium chloride

1.12 80/100 grade bitumen specifies

(a) 8 to 10 mm penetration at 25°C, for a 5 second water bath

(b) 80 to 100 mm penetration at 25°C, for a 5 second water bath

(c) 8 to 10 cm penetration at 25°C, for a 5 second water bath

(d) 80 to 100 cm penetration at 25°C, for a 5 second water bath

1.13 Steel is prevented from rusting through galvanization by addition of thin film of

(a) Magnesium (b) Chromium (c) Zinc (d) Nickel

1.14 The oil used to thin and distribute the paints is known as

(a) Base (b) Vehicle (c) Solvent (d) None of the above

1.15 Which of the following is used as a solvent in varnish paints

(a) Shellac (b) Copel (c) Amber (d) Turpentine

Answers of Multiple Choice Questions

Answers of Multiple Choice Questions
1.1 (b), 1.2 (c), 1.3 (c), 1.4 (b), 1.5 (a), 1.6 (c), 1.7 (b), 1.8 (d), 1.9 (c), 1.10 (c), 1.11 (b), 1.12 (a), 1.13 (c), 1.14 (c), 1.15 (d)

Short and Long Answer Type Questions

1.1 Name the Bogues compound responsible for the early strength of cement.

1.2 Describe the temperature at which clinkers are formed in the dry manufacturing process of cement.

1.3 Which type of cement offers a lower early gain of strength?

1.4 Give the advantages of M sand over river sand.

1.5 Give the significance of curing in strength gain of concrete.

1.6 Describe raw materials used in the manufacturing process of glass.

1.7 Name the common forms of timber.

1.8 Name the types and grades of reinforcement bars.

Numerical Problems

1.1 For preparing a M 20 grade of nominal mix plain cement concrete, the weight of aggregate and water is given as 250 kg and 30 kg, respectively, for a 50 kg bag of cement. Assuming that the ratio of fine and coarse aggregate is 1:2, and density of concrete is 2400 kg/m^3. Find the weight of materials required to prepare a concrete slab of 2.5 m × 1.5 m × 0.5 m.

1.2 1.0 kg of aggregates were sieved to identify the particle size distribution. The results of particles retained on different sieves are given as follows:

Sieve designation	10 mm	4.75 mm	2.36 mm	1.18 mm	600 μm	300 μm	150 μm
Weight retained	0 g	10 g	120 g	170 g	400 g	180 g	20 g

On the basis of given particle size distribution, classify the aggregates as per IS 383.

PRACTICAL

1. Identify ten different engineering materials which you can observe in your everyday constructions. You may adopt examples from your home, college, nearby buildings, roads, etc.
2. Along with each engineering material mention the application under in which the material was used.

3. On the basis of application list down at least three desirable properties of that material. Also suggest an alternative engineering material which may also satisfy the desirable properties.

KNOW MORE

Sustainability

The unit covers the technical performance of civil engineering materials, which is essential for their satisfactory application. However, in reality there are two more parameters which govern the satisfactory application of any engineering material – economy and environment. When an engineering material meets all three criteria's, i.e., environment friendly, technically suitable and commercially viable, it is known as a sustainable material. Engineering materials can also be classified on the basis of their contribution towards sustainability parameters.

Following provides the different types of engineering materials and their contribution towards sustainability:

Technically satisfactory material (technical)

Economic material (economy)

Eco-friendly material (environment)

Environmentally viable materials (economy and environment)

Equitable materials (economy and technical)

Bearable materials (environment and technical)

Sustainable materials (economy, environment and technical)

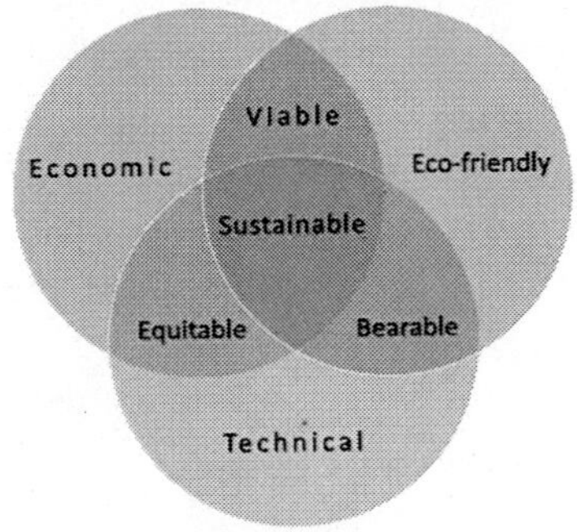

Fig: Types of engineering materials in terms of sustainability

It should be kept in mind that sustainability is a relative terminology. For example, lack of fly ash in north eastern parts of India can make Portland pozzolana cement (PPC) a costly cement; while PPC presents is a sustainable material in many parts of India. Similarly, recent advancement may also shift the classification of materials. It is said that for many parts of the world that geopolymer is a more sustainable solution over PPC. One should have an understanding of sustainability so that suitable selection of materials can be made.

Sustainable development

Development (infrastructure, economic, social, etc.) using sustainable materials is known as sustainable development. Sustainable development is an ongoing international collective effort, which formally received recognition in 2015 by the United Nations. One of the significant components of this effort is to make cities and human settlements inclusive, safe, resilient, and sustainable (SDG 11). To realize this goal, researchers and relevant stakeholders have investigated and demonstrated a wide range of sustainable practices. One such widely recognized practice is the use of sustainable construction materials. Over the years, many wastes and alternative resources have been identified for their potential as sustainable construction materials. These new materials have different impacts on the properties of structures in terms of environment, technical performance, economy, service life, and human comfort. Limited understanding of sustainable construction materials and their compound effects on the property of structures is a significant barrier for the industry to move towards sustainable construction. In India, organizations like BMTPC, CBRI, NITI Aayog, DST, AICTE and several others are actively engaged in creating awareness towards the development and application of sustainable engineering materials.

Future trends for civil engineering material

Over the year, Indian construction industry has given a preferential treatment towards technically satisfactory and economically feasible materials. Environment was often overlooked, as it is a collective effort and industry avoided bearing its cost. Recognizing this challenge Government of India has started making changes which promote the application of sustainable engineering materials. These changes vary for different materials, like, mandating the use of fly ash, restricting the use of natural sand, banning certain application of lead and giving preferential treatment for low carbon alternatives in government tenders. As a result of these changes construction industry has started transitioning towards sustainable engineering materials, and there is growing demand for engineers with the knowledge for such materials.

Any individual moving in the direction of sustainable engineering materials should be aware about the following *Life cycle assessment (LCA)* and *life cycle cost (LCC)*. LCA is used measure the negative impact of any engineering material on the environment. The negative impact is represented in terms of LCC. LCA or LCC for a material can be from cradle to gate, i.e., when the material is used, or cradle to grave, i.e., when the material has been fully used and disposed. LCC is typically measured in terms of equivalent carbon emission and equivalent embodied energy, however other parameters may also be included like sulphur, nitrogen and CFC emissions.

REFERENCES AND SUGGESTED READINGS

1. S. S. Bhavikatti, *Building Material and Construction,* Vikas Publishing House Pvt. Ltd., 2020.
2. N. E. Dowling, *Mechanical Behavior of Materials,* Pearson, 2017.
3. S. C. Rangwala and K. B. Dalal, *Engineering Materials,* Charotar Publishing House Pvt. Ltd., 2019.
4. G. Singh, *Building Construction and Material*, Rajsons Publication Pvt. Ltd., 2022.
5. C. B. Kukreja, K. Kishore and R. Chawla, *Material Testing Laboratory Manual (For Quality Control)*,Standard Publishers Distributors, 2018.
6. V. Raghavan, *Materials Science and Engineering*, PHI Learning Pvt. Ltd., 2021.
7. I. P. Singh, S. Chander and R. K. Prasad, *Materials Science and Engineering*, Jain Brothers, 2020.
8. R. K. Sharma, *Testing of Construction Materials*, Wiley, 2019.
9. J. P. Patel and G. H. Upadhyay, *Material Science*, Atul Prakashan, 2001.

10. M. L. Gambhir and N. Jamwal, *Building and Construction Materials* , McGraw Hill education, 2014.

11. S. K. Duggal, Building Materials, New Age International (P) Limited, Publishers, 2019.

12. N. Subramanian, *Building Materials Testing and Sustainability*, Oxford University Press, 2019.

13. PK Mehta, PJ Monteiro, *Concrete: microstructure, properties, and materials*. McGraw-Hill Education; 2014.

Dynamic QR Code for Further Reading

2 Emerging Civil Engineering Materials

UNIT SPECIFICS

Through this unit, we have discussed the following aspects:

- *Emerging materials with potential for application in civil engineering;*
- *Physical attributes of typical structural and geotechnical building materials;*
- *Process of preparing manufacturing materials, along with directions for modifications for desired material properties*

The different types of civil engineering materials are discussed to foster an appreciation towards alternative building materials and the various properties exhibited by them, to encourage the efficient use of civil engineering materials.

Along with a large number of multiple-choice questions and questions with short and long answers, the unit also includes a list of references and suggested readings that one can use as practice and grow their knowledge on civil engineering materials.

RATIONALE

This unit on emerging civil engineering materials helps students to familiarize with the recent advances in civil engineering materials. The unit highlights several emerging civil engineering materials such geotextile, graphene and carbon composites. The discussion is followed by some problems related to civil engineering materials, which will further help in developing a clear idea of the concerned topics on the subject.

A large number of emerging materials are getting introduced in civil engineering applications. A good grasp of emerging materials is necessary for adapting construction practices with changing times. Without this information, it will be challenging to resolve various evolving needs. An overview of all the emerging civil engineering materials will help civil engineers and other stakeholders to appreciate alternative resources and encourage their application.

PRE-REQUISITES

Chemistry: Chemical equations and commonly used notations (Class XII)

UNIT OUTCOMES

As outcome of this unit, one should be able to:

U2-O1: Understand engineering usage of various materials.

U2-O2: Study the ingredients and composition of various materials.

U2-O3: Comprehend the manufacturing process of various materials

U2-O4: Apply appropriate material for engineering purposes.

Unit-1 Outcomes	EXPECTED MAPPING WITH COURSE OUTCOMES (1- Weak Correlation; 2- Medium correlation; 3- Strong Correlation)							
	CO-1	CO-2	CO-3	CO-4	CO-5	CO-6	CO-7	CO-8
U2-O1	2	-	2	2	-	3	2	2
U2-O2	1	-	1	-	-	-	2	2
U2-O3	-	-	-	-	-	-	1	1
U2-O4	3	1	2	2	2	2	3	2

2.1 INTRODUCTION

Recent advances in the field of material science have led to a development of a large number of materials with potential of application in civil engineering. These emerging materials are identifying both conventional and novel applications in the field of civil engineering. With the introduction of these engineering materials, engineers are able to resolve challenges, which previously seemed unthinkable. Examples of emerging civil engineering includes plastic, which can drastically reduce the weight of construction. Other types of emerging civil engineering materials include ceramics, refractories, and geotextiles. It is important that stakeholders are aware of the emerging materials so that they may adopt these materials as per their need.

2.2 CERAMICS

Ceramics are defined as a class of inorganic, non-metallic solids that are subjected to high temperature in manufacture and/or use. According to the American Ceramic Society, the term "ceramic" has been derived from Greek word "Keramikos" meaning pottery, and refers to the materials developed from fired clay. Ceramics are relatively cheaper engineering materials due to the abundance of its raw materials. Hence, they are used in a wide number of industries like refractory materials, tiles, glasses, porcelain cement and concrete manufacturing, chemical porcelain, porcelain enamels, insulating and semiconducting materials, white wares (tiles and insulators) and stone wares (glazed pipes and roofing tiles).

Commonly used ceramics are composed of oxides, carbides, and nitrides. Silicides, borides, phosphides, tellurides, and selenides also are used to produce ceramics. Ceramic is generally produced at high temperatures, and the resulting material typically shows heat resistance or refractory properties. Ceramics are typically processed by powder metallurgy process, as depicted in Fig. 2.1, which involves the conversion of raw materials into powdered form, blending into a

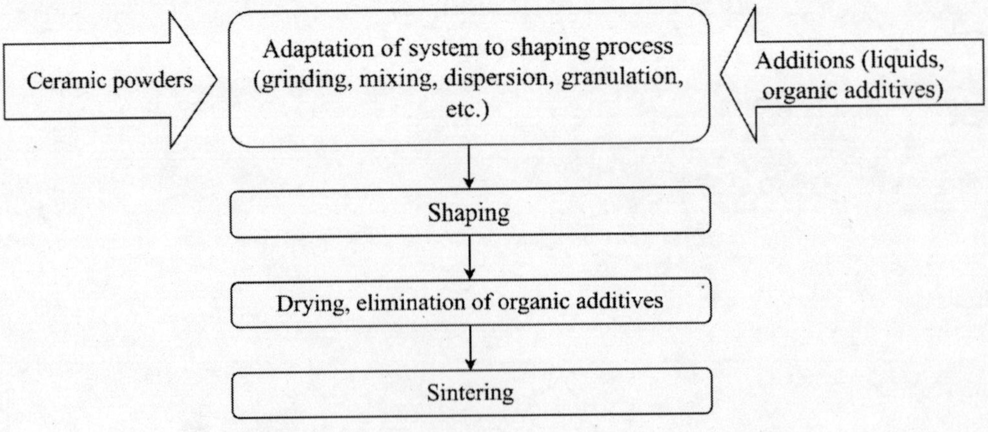

Fig. 2.1: A typical ceramic manufacturing process

homogeneous mixture, then transferring the mixture in a mould and applying pressure to consolidate it, and eventually performing sintering (heat application) on it. After sintering, machining may be performed for better dimensional accuracy, by means of laser beams or electron beams (the use of cutting tools may break the ceramic, due to its brittle nature).

Some of the properties related to ceramics are as follows:

- They are poor conductors of heat and electricity due to lack of free electrons.
- They are highly corrosion resistant.
- They have high melting point temperatures.
- They are hard and tough.
- They have low ductility.
- They have high strength and hardness at high temperatures.
- They have high compressive strength and low density, hence, can be used to make thinner and lighter products.
- They are weak in tension due to their brittle nature.
- They have poor impact resistance.
- They have high dimensional stability.

2.3 REFRACTORIES

Refractories are those materials which can resist the effect of high temperatures and deliver the desired performance. As per AFNOR (French Standardization Association), a material (other than metals and alloys) is said to have refractory property, if it exhibits a pyroscopic resistance equivalent to at least 1500 °C. Refractory materials are suitable for use in high-temperature areas like lining blast furnaces, boilers, nuclear reactors, and crucibles. It is used to manufacture spark plug insulators, cutting tool bits, and electrical device insulators.

Refractories can be produced from different materials like fire clay, alumina and slags. In general, refractories are manufactured by the processes depicted in Fig. 2.2. Raw materials are crushed and screened to obtain the raw ingredients with desired size. After screening, the raw materials are mixed in desired proportions to obtain a homogenous blend. The mix, in wet state, is given the desired shape or form and dried to remove excess moisture. Then, once the mould is prepared, firing is done in continuous kilns to form ceramic welded bonds. After firing, the product is cooled and then given a proper finish before using them.

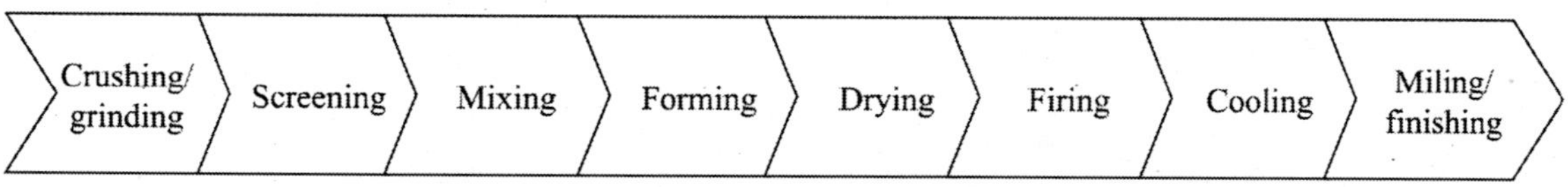

Fig. 2.2: Typical process for manufacturing refractories

Type of refractories:

Refractory materials are mainly of three types on the basis of their chemical behaviour, and are described as follows:

Acidic refractories: Remain unaffected by the presence of acid slags but are weak when exposed to alkaline slags. These are based on alumina-silica compositions i.e., alumina, silica and zirconia refractories.

Basic refractories: Remain unaffected by the presence of alkaline slags but are weak when exposed to acid slags. These are based on magnesia compositions, i.e., magnesite and dolomite.

Neutral refractories: Remain unaffected by the presence of acid and alkaline slags. They are typically made of silicon carbide, chromite, and carbon.

Another classification for the refractories is based on the composition of raw ingredients used, and can be described as follows:

Fireclay refractories: Contain aluminosilicate with silica content less than 80% and alumina content less than 44%. They are relatively cheap and have applications in kilns, stoves, and furnaces, iron and steel industries, cement industries, pottery, and glass industries. In some cases, fire clay refractories are made highly porous to reduce thermal conductivity and resulting heat losses in furnace.

Siliceous refractories: Contain at least 93% silica. They have good mechanical properties at temperatures near their fusion temperatures. These are mostly used in metallurgical furnaces and glass tanks.

High alumina refractories: Contain at least 45% alumina. They are used in the hearth and shaft of blast furnace, cement, lime, and ceramic kilns.

Magnesite refractories: Contain at least 85% magnesia. These are used to manufacture basic bricks, which are used to resist basic slag effects during steel melting, hence used in extractive metallurgy.

Chromite refractories: Contain 15 to 35% chromium oxide and around 50% magnesia. These acidic bricks are used to withstand corrosive flux and high-temperature glasses, hence used in steel melting.

Zirconia refractories: Contain high zirconia content, typically above 60%. High heat resistance makes them retain the strength up to 1500 °C. Hence, they can be used as high-temperature constructional materials for furnaces.

2.4 PLASTIC

Plastic is a synthetic material of long molecular chains made from different monomers. It can be converted into our desired shape when in melted form and then transformed into a rigid or slightly elastic form, in the presence of catalysts. It came from the Greek word *"plastikos"* which means *"capable to be shaped or moulded"*. Low density, ease of fabrication, low thermal and electrical conductivity, pest-resistant, transparency as per requirement, and low unit costs during mass production makes plastic an attractive option to engineers as an engineering material.

Plastics can be classified into two broad classifications based on the manufacturing method, thermoplastics, and thermosets, as shown in Fig. 2.3. Thermoplastics are plastics which don't undergo chemical changes in their compositions when heated, hence they can be reheated and remoulded several times. Hence, they can be used as recyclable materials. These are generally solid at room temperature but become soft and fluidified when heated. Due to its fluid and adhesive nature, it adheres well to metals, hence giving an aesthetic finish. However, they are costlier than their thermosetting counterparts. Examples include polypropylene, PVC, polyethylene, polystyrene, etc. Thermosets are usually resinous at room temperatures but solidify when heated in presence of additives. These plastics form permanent chemical bonds, called crosslinks, which inhibit them from returning to the melted state. Hence, thermosets can be heated and moulded only once. If heated further, they decompose rather than melting to change shape. Thermosets can resist chemical attacks better than thermoplastics. Examples include polyamides, and epoxy resins.

Thermoplastics

Thermosetting plastics

Fig. 2.3: Types of plastics

(Source: https://themechanicalengineering.com/difference-between-thermoplastic-and-thermosetting-plastic-with-pdf/)

The major drawback of most plastics is their non-biodegradable nature. Hence, when littered around, they affect water bodies by clogging waterways, choking animals when unintentionally consumed, causing cancer when food is directly eaten from plastic packages, and both the manufacturing and incineration of plastics produce harmful gases. It should be noted that noticeable work has been carried out in the direction of environment friendly plastics, like biodegradable plastics (those which get degraded when exposed to sunlight, bacteria, enzymes, water, etc.) and bioplastics (biodegradable plastics made entirely of cellulose and starch). Another drawback of plastics is that, due to their relatively low stiffness, plastics are still not used as load-bearing members, hence, their applications are primarily limited to waterproofing, packaging, flooring, and pipe and sanitary appliance manufacturing purposes.

2.5 ALUMINIUM

Aluminium is silver-white in colour and has a brittle metallic shine. Aluminium has low thermal conductivity and is an excellent conductor of electricity. It is highly light, soft, robust, and durable. Although aluminium cannot be soldered, it may be riveted and welded. It can be tempered at 350°C. The tensile strength is usually around 117.2 N/mm^2 in cast form and 241.3 N/mm^2 when formed into wires. The melting point of aluminium is around 660 °C.

Aluminium shows better resistance to harsh environments, as compared to steel. Aluminium is resistant to nitric acid, but concentrated sulfuric acid and hydrochloric acid dissolve it slowly. Sulphur, carbonic acid, carbonic oxide, vinegar, seawater, etc. have little effect on it at room temperature, while caustic alkalis quickly erode it.

Bauxite ($Al_2O_3.2H_2O$) is the primary raw material used to manufacture aluminium on a large scale, containing aluminium oxide as the main constituent and iron oxide, silica, and titania as minor constituents in varying proportions. Other aluminium ores include corundum, kaolin, and cryolite.

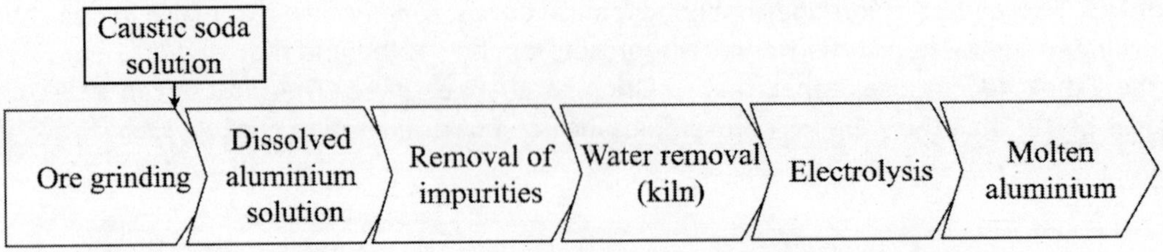

Fig. 2.4: Flow diagram for the extraction of aluminium

Aluminium can be produced by refining the ore using Bayer's technique, followed by Hall Heroult's process for converting the refined ore into aluminium. Hall Heroult's is a two-stage process. In the first stage, roasting, grinding, heating (with sodium hydrate), and filtering turns bauxite into alumina. The hydrate precipitates after many hours of stirring, it is then separated, cleaned, and calcined at about 1000 °C. Alumina is then electrolyzed in a molten crystolite bath to extract aluminium (Fig. 2.5).

Uses for Aluminium:

Aluminium is very soft in its pure form and not preferred for direct use in structures. By alloying aluminium with copper, nickel, zinc, and manganese, satisfactory characteristics can be obtained. Alloys of aluminium work well for manufacturing corrugated roofing sheets, shop railings, and door and window frames. Aluminium is also used in form of sheets to prevent decay, for a wide range of applications like internal combustion engines, aircraft, kitchenware, and packaging for goods like chocolate and medications.

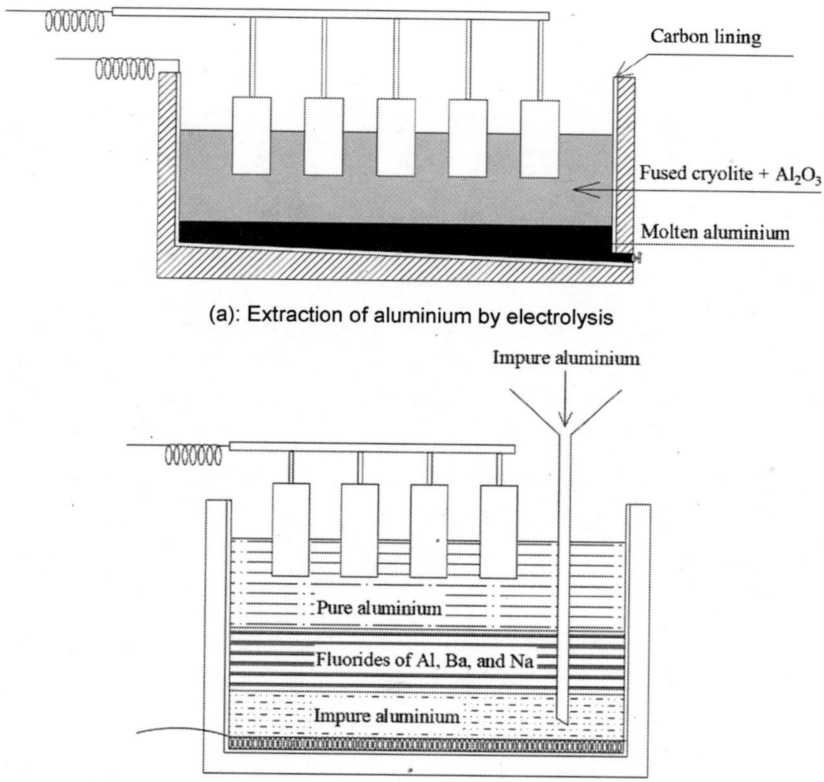

(a): Extraction of aluminium by electrolysis

(b) Hoope's cell for refining aluminium

Fig. 2.5: Process for aluminium extraction

Alloys of Aluminium:

Aluminium is frequently alloyed with copper, silicon, magnesium, or zinc to improve its mechanical qualities. The metals manganese, lead, nickel, chromium, titanium, and beryllium may also be present in some aluminium alloys. Aluminium alloys can be divided into two categories: cast alloys, which are moulded through casting, and wrought alloys, which are mechanically worked into various shapes. Cast alloys are typically binary alloys made of silicon, copper, and occasionally magnesium. Copper, magnesium, silicon, and manganese are components of wrought alloys, which combine with aluminium to generate precipitation-hardening alloys. Some of the well-recognized aluminium alloys are as follows:

Duralumin: Duralumin typically contains of 3.0 - 5.0 % copper, 0.50 - 2.0 % magnesium, and 0.3 - 1.0 % manganese, with trac number of impurities in the form of iron and silica. Duralumin is very light, with a relative density of about 2.80. The tensile strength is comparable to mild steel, when rolled and heat treated. It has a high level of corrosion resistance. Duralumin can be drawn into wires and sheets. It can be produced into a variety of structural shapes for usage in buildings.

Magnalium: Magnalium is lighter than pure aluminium, with specific gravity of about 1.90. Magnalium contains about 5 - 6 % magnesium. It is used as a deoxidizer in copper smelting operations.

Aldural: When duralumin is coated with aluminium, it is called "aldural" and has better corrosion-resisting properties.

2.6 COPPER

Copper is a red-orange coloured metal with a metallic lustre. It has high thermal and electrical conductivity, as compared to most metals. It is soft, malleable and ductile. In its pure form, copper is widely used for electrical wirings. Copper is used as alloys in structural use.

Copper is extracted from ores containing copper pyrite, such as chalcopyrite ($CuFeS_2$), malachite ($CuCO_3 + Cu(OH)_2$), and copper glance (Cu_2S). Smelting is the main method used for obtaining copper. in this process, the copper ore is blended with coke and silica, which separates most of the iron and sulphur compounds. The resulting crude copper is known as blister copper, which is then refined either using the electrolysis or the reverberatory process.

Alloys of Copper:

Zinc, lead, and tin are most common elements to be used as alloying material in copper. Most common alloys are brass and bronze.

Brass: It contains zinc as an alloying material, typically in the range of 10-40%. The colour of brass is silver-white but with an increase in copper content, the colour becomes red. Brass is more ductile and stronger than its components, i.e., copper and zinc. Forging and rolling improves at higher concentrations of zinc. Brass for cold work has copper to zinc ratio between 2:1 and 3:1. Other metals are also added in small quantities to improve the properties of brass. Addition of tin reduces ductility but increases hardness. With the addition of lead, the ductility decreases but its ability to turn and polish increases. With the addition of aluminium, the strength increases but the ductility decreases. Lead induces flowability of brass.

Bronze: Bronze is a metal alloy made of copper, tin, and one or more other metals. The bronze used to make coins is known as coinage bronze, which typically consists of tin (4%), and zinc (1%) are alloyed. Gun metal, which typically contains 10% tin, and 2% zinc, is used to make valves and bearings. Bell metal is created by combining copper with tin, and is used to create appliances, fixtures, and electrical goods.

Phosphor bronze: A small amount (0.01-0.5%) of phosphorous serves as a deoxidizer in the copper and tin alloy. The resulting alloy is known as phosphor bronze. Up to certain quantity phosphorous enhances hardness and brittleness. If too much phosphorus is added, the product loses its usefulness.

Lead bronze: Copper, tin, phosphorous (less than 1%), and lead (around 3%) are all present in lead bronze. It works best when used to create bearings. Bronze segregates when lead content exceeds about 4%, creating soft areas in the hard matrix.

Manganese bronze: It is used in marine engines and propeller blades as it has a high strength of nearly 460 N/mm^2. It contains a small percentage of tin, iron, manganese and lead.

2.7 TIN

Tin is primarily extracted from cassiterite (SnO_2), an oxide of tin. Tin is extracted from the ore by crushing, roasting, smelting and refining, in a way similar to that of copper.

Properties of Tin: Tin is a beautiful, silvery-white metal that is incredibly malleable. It can be cut with a knife because it is very soft. It has a specific gravity of about 7.3, and melts at around 230 °C. Tin is ductile with strong corrosion resistance.

Uses: Tin coated sheets are used to make cans, utensils, and furnace pipes. Lead tin alloy coated sheets are used for roofing. Tin is also used for making bronze and other alloys.

2.8 NICKEL

Nickel is a silvery-white metal with a slight golden tinge. Nickel is generally extracted from pyrite or silicate ores.

Properties: Nickel polishes well and does not tarnish or corrode in dry air when exposed to normal temperatures. Nickel has a specific gravity of about 8.9 and melting point of about 1500 °C. Nickel is hard, malleable and ductile.

Uses: Nickel is frequently used as an alloying material and as an electrodeposited coating to prevent corrosion on steel. In addition, nickel is used to create alloys with the metals iron, copper, and chromium.

2.9 ACOUSTICAL MATERIAL

Engineering structures are typically designed to shield the user from various factors like rain, sunlight, wind, etc., but often overlook factors like noise or sound. When the intensity of sound is higher than the comfort limit, it is termed as noise. For example, extreme horns and heavy machine noises can cause discomfort. The noise can be managed by appropriate material known as *"acoustical material"*. Acoustical material lowers the intensity of noise (sound) so that it falls within comfort level. Typical application areas of acoustical materials can include auditorium, movie theatre, studio, recreation centre, entertainment venue or college reading room.

There are two main applications for acoustical materials: soundproofing, which prevents noise generated outside of a specific space from entering the interior, and sound absorption, which reduces the noise generated inside a space. A school might build a specific wall to segregate the music room from the normal classroom next door as one example of soundproofing. Another example could be a machine shop that might install sound-absorbing barriers to prevent and take in the acoustic energy of a noisy air compressor. Acoustical materials can also be used to manage both internal and external sounds. For example, theatres require acoustic materials to prevent internal sounds from escaping and disturbing outside individuals, as well as external sound from entering and interfering with listeners' experience.

Acoustical materials can be made using a variety of materials like foam, textile and metals. Recent study shows that the shape of materials can also be modified to generate metamaterials with sound proofing capabilities.

Measures of acoustical material effectiveness:

Manufacturers measure and define various qualities of acoustical materials, to estimate their effectiveness. The most typical of these are:

Absorption coefficient

Acoustic impedance

Noise reduction coefficient

A-weighted sound level scale

Absorption coefficient of a material is defined as the ratio of energy absorbed by the engineering material to the incident energy of the sound. Absorption coefficient can be defined for any frequency range of sound.

Acoustic impedance of a material is defined as the product of its density and acoustic velocity. Frequency affects an acoustic material's capacity to absorb sound energy, with most materials effective at reducing mid-to-high frequencies than lower ones.

Noise reduction coefficient refers to the measurement of a material's average sound absorption coefficient at frequencies of 250, 500, 1000, and 2000 Hz that can be used to assess how well a material typically absorbs noise. The noise reduction coefficient does not cover the lower base range frequencies, which are often the most problematic.

A-weighted sound level scale is used to measure the sound in human-occupied environments. This scale lessens the impact of high and low frequencies to match the range of human ear response.

Table 2.1 below illustrates the correlation between decibel scale and the equivalent energy. Remember that the energy level increases by an order of magnitude for every 10 dB shift by a factor of ten.

Table 2.1: Decibel scale of sound pressure

Reference sounds	Decibels (Reference level = 20 μPa)	Energy Level
Hearing threshold	0	1
	10	10
Tree leaves fluttering	20	10^2
Whisper in an ear	30	10^3
	40	10^4
	50	10^5
Normal speech	60	10^6
	70	10^7
	80	10^8
	90	10^9
Cars/vehicles from nearby	100	10^{10}
	110	10^{11}
Airplane taking off	120	10^{12}

Sound control using acoustic material:

Different mechanisms are used to control sound and noise, i.e., absorption, damping, decoupling, distance, and adding mass. These mechanisms are used for designing various acoustic materials, of which the two prominently used materials are sound proofing materials and absorbers.

Sound proofing materials: Sound proofing refers to the concept of preventing sound transmission from one space to other. This is typically achieved by either reflecting or absorbing the incident sound. Sound proofing materials can be prepared by several techniques, some of which are as follows:

- By placing a layer of viscous damping adhesive between two layers or by using sound-dampening liquid coatings on surface, sound can be dampened. Acoustic caulk or other sound-dampening sealants are available for this purpose.
- By installing thick vinyl sheeting behind sheetrock, prevents the energy from being transferred to the more solid inner walls. The product under sheetrock also improves the building's fire resistance.
- By using high density materials like lead-sheet, that effectively reduces sound transmission through walls. The lead is often attached to foam on both sides before being glued to sheetrock.
- By mechanically decoupling interior walls from external construction. Typically, sound isolation clips are used. These clips, which consist of rubber-fitted mounts that hold nailing strips, offer a practical way of isolating a space from outside disturbances. Sound isolation clips are solely intended to operate on airborne sound, and are not effective for, noise passing through buildings, pipes, etc. won't be muted.

Apart from buildings, soundproofing materials are used in commonly used in aircraft to shield the crew and passengers from engine noise and high-frequency flight noise.

Absorbers: Absorber or sound-absorbing materials absorb the sound and lower the intensity of both transmitted sound and reflected sound. Absorbers can be prepared by several techniques, some of which are as follows:

- By using acoustic foam. Acoustic foam often has an open-cell structure because closed-cell structures generally reflect sound rather than absorb it. Melamine foam is an example of acoustic foam. Acoustic foam often uses different geometries which further help in reducing the spread of sound.
- By using hanging acoustical baffles and dividers. They might be soundproof, movable room dividers or flat permanently placed partitions.
- By using acoustical blankets. They as typically used in industrial settings to set up noise absorption around machines. These can be used as either short-term or long-term noise absorbers. Similar materials can also be used to cover pipes, reducing the amount of noise that passes through them.

Soundproofing vs Sound Absorption: The idea that soundproofing and sound absorption are interchangeable is a common misconception. In reality, they are not the same. Incident sound gets either transmitted, absorbed or reflected. Sound proofing element focuses on reducing the intensity of transmitted sound either by reflection or absorption. On the other hand, sound absorbing element focuses on reducing both transmitted and reflected intensity of sound. Sound proofing element may reflect the noise and create discomfort inside structure. Sound absorption can prevent sound from echoing. For instance, in a noisy restaurant with hard surfaces typical of modern trendy restaurants' industrial-influenced decorations, guests' conversations and the sound of clattering cutlery may generate a cacophony of noise. Thus sound-absorbing materials are preferred in such spaces as they reduce the resonance and reverberation of sound waves inside a room or space along with soundproofing the space.

2.10 GEOTEXTILES

A geotextile is a synthetic permeable fabric, normally used to improve the qualities of the soil. It can be used as a separation layer for different types of materials, while permitting sufficient liquid flow between them. The porosity and permeability properties of geotextiles are also used for filtration. Due to better internal friction with the soil grains, geotextiles also provide reinforcement or tensile strength to the soil. As a result, geotextile have become a popular choice in pavement construction, drainage systems and soil stabilization. At present, geotextiles have been created, manufactured, and used for soil modification and improvement in several parts of the world, prominently in the US, Japan, and western Europe.

Properties of geotextiles:

For geotextiles to be utilised in the civil engineering field, they must have several qualities, such as strength and flexibility. It should have elongation properties without rupture and should have high abrasion resistance and resistance against microbial attacks. Since geotextiles are typically used in large and open spaces, they should be able to have a wide fabric. Furthermore, the geotextile should be porous and permeable to support the flow of water.

Types of geotextiles:

Geotextiles are designed using permeable fabrics made from natural sources, like coir, or synthetic sources, like propylene. They are divided into 3 categories on the basis of their preparation:

- Woven Fabric Geotextiles
- Non-Woven Geotextiles
- Knitted Geotextiles

Woven fabric geotextiles: These are generally woven using methods that are comparable to those used to weave regular garment textiles. Lateral and longitudinal sets of threads or yarns occur in this type's distinctive pattern. Warp refers to the longitudinal yarn that runs parallel to the length, and weft refers to the lateral yarns that run perpendicular to length.

Non-woven geotextile: Short staple fibres or continuous filament yarns are used to make non-woven geotextiles. Fibres can be joined together mechanically, chemically, thermally, or a combination of these methods.

Fig. 2.6: Non-Woven Geotextile

(Source: www.indiamart.com/proddetail/non-woven-geotextile-fabric-21528352230.html)

Knitted Geotextile: In this textile, different yarn loops are interlinked together. It is done by using needles and continuous threads. The procedure of weaving a number of yarn loops together results in knitted geotextiles. All knitted geosynthetics are created by combining the knitting technique with another geosynthetic production technique, such as weaving.

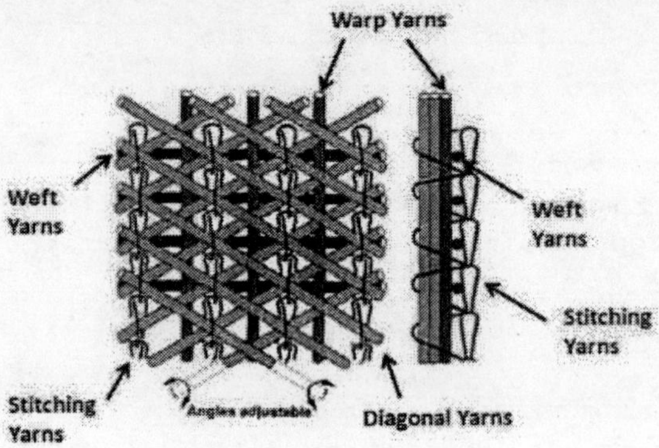

Fig. 2.7: Knitted Geotextile

(Source: https://theconstructor.org/building/geotextiles-types-functions-uses/1163/)

Functions of Geotextiles:

Functions of geotextiles are shown in Fig. 2.8, and can be described as follows:

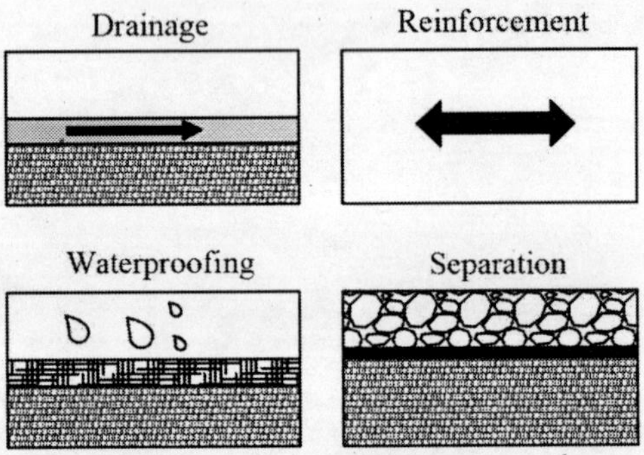

Fig. 2.8: Functions of Geotextiles

Separation: The intermixing of two adjacent soils is avoided by the separation property of a geotextile. It is responsible for the load transfer and drainage properties while separating the different courses of a road. It is used for the separation of sand drains and geomembranes, in railroads, stone bases and landfills.

Filtration and drainage: With the properties of permeability and porosity, geotextiles can control the liquid flow and can be used for filtration and drainage. Filtration can be used during soil stabilization for improving ground water recharge, and drainage can be used for movement of water between different layers of pavement.

Reinforcement: Like steel in concrete, geotextile acts as reinforcement and improve the tensile strength. The following mechanisms explain how the addition of geotextile increases soil strength:

- Lateral constraint caused by friction at the interface between the geotextile and the soil or aggregate.
- Membrane-style wheel load support by improving the shear strength of the soil.

Waterproofing: Geotextile absorbs asphalt or bitumen to create a waterproof barrier. Geotextile in some cases is used between old and new asphalt during repair of pavements, and creates a water proof barrier for the protection of pavements.

Applications of Geotextile in Construction:

It can be used in different construction projects. A few of the prominent applications are:

Road Work: In road construction, it is used as a separator or sealant. It also acts as reinforcement to improve tensile strength. Since it serves as a layer for rapid dewatering of the roadbed, it acts as a separator and also maintains its permeability.

Fig. 2.9: Use of Geotextile in road construction

(Source: https://theconstructor.org/building/geotextiles-types-functions-uses/1163/)

Railway Works: It separates the subsoil without disturbing groundwater flow. Each layer is provided as cloth, this also helps in reducing the sidewise movement from generated from the shocks and vibration of trains.

Fig. 2.10: Use of geotextile in railways

(Source: https://theconstructor.org/building/geotextiles-types-functions-uses/1163/)

Agriculture: It is typically employed to stabilize mud. Nonwoven fabrics are folded by overlapping, to include a pipe or a mound of grit in muddy routes and trails that are used by cattle or light traffic.

Drainage: The drainage property of geotextiles is used to drain water through roads with the use of granular materials. It serves as the filtering mechanism for drainage in reservoirs, on roads and highways, in earth dams, behind retaining walls, in deep drainage trenches, and in agriculture.

Fig. 2.11: Use of Geotextile in drain construction

(Source: https://theconstructor.org/building/geotextiles-types-functions-uses/1163/)

River, canals and coastal works: River banks are shielded from erosion caused by water currents by overlapping them with geotextiles. It also serves as a filter when utilised with either natural or man-made embankments.

Fig. 2.12: Use of Geotextile in canal construction

(Source: https://theconstructor.org/building/geotextiles-types-functions-uses/1163/)

2.11 RUBBER

Rubber, also referred to as elastomer, is a high-density polymer formed of spring like polymeric chains, as shown in Fig. 2.13. As polymers are in the form of coils or springs, they easily elongate (under tension) or shorten (under tension) without breaking of bonds. Furthermore, the spring like shape also helps in recovering the original shape of the rubber when external loads are removed. Rubber is mainly preferred for its property of experiencing significant deformations under load and recovering to its original shapes and dimensions after its removal.

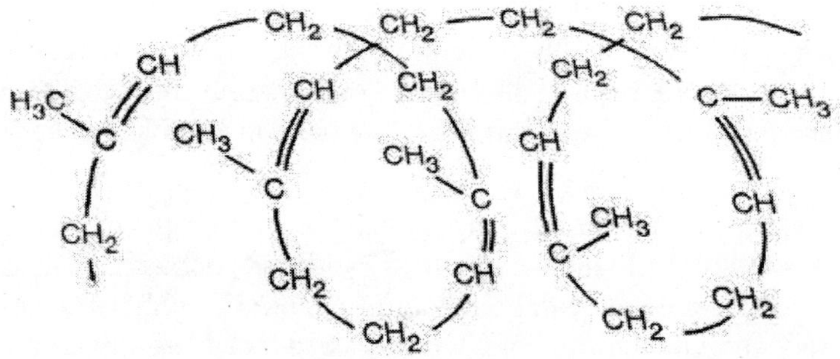

Fig. 2.13: Schematic representation of coiled elastomer

(Source: www.ques10.com/p/32856/what-is-natural-rubber-explain-synthesis-and-dra-1/?)

Classification:

Rubber can be classified as natural rubber or synthetic rubber.

Natural rubber: The primary component of natural rubber is latex (isoprene). Polymerization of latex results in a long-coiled chain of polyisoprene (Fig. 2.14).

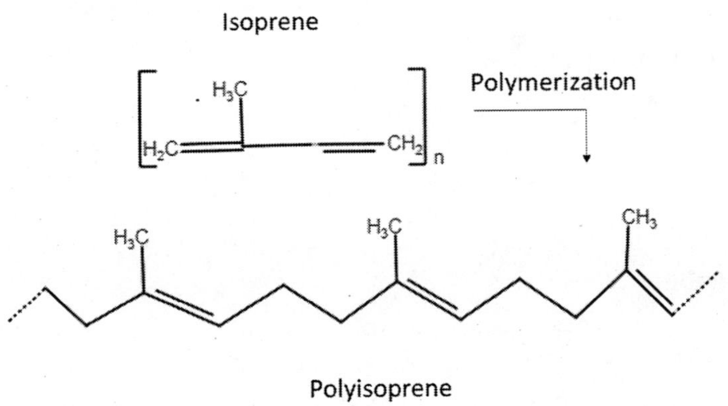

Fig. 2.14: Polyisoprene (Natural Rubber)

Latex is generated from sap found in the cells of many plant sections. Latex is extracted by creating slits in the rubber tree's bark and drinking out the sap. Gathered latex is diluted to contain 15-20% of rubber, and then filtered to get rid of any impurities. After that, acetic acid or formic acid is added to the rubber to coagulate the sap in form of a rubber mass. From coagulated rubber, crude rubber is made which is further treated to produce commercial rubber.

Crepe Rubber – Coagulation of rubber in the presence of a retarder, such as sodium bisulphite, results in the formation of crepe rubber. Most of the serum is then squeezed out as it is passed between the creping machine's rollers, creating a sheet that resembles crepe paper.

Smoked Rubber – It is made by adding diluted latex to tanks with vertical slots in the sidewalls. Diluted formic or acetic acid is added to the latex and stirred. After being placed in the grooves, the partition plates are left there for roughly 16 hours. Then the resulting rubber mass is placed in rollers with smaller clearances. The resulting slabs are removed and put through a sequence of rollers with progressively smaller clearances. The sheets are maintained in a smoke house with a temperature of 40 to 50 °C for four days.

Synthetic rubber – Natural rubber and thermoplastic vinyl high polymers serve as the basis for synthetic rubber. There are an infinite number of synthetic rubbers that could exist. This is true because any straight-chain polymer can be tailored to generate rubber-like characteristics. Natural rubber is inexpensive and widely available, which has reduced the demand for synthetic rubber. Materials created synthetically to resemble rubber are referred as elastomers.

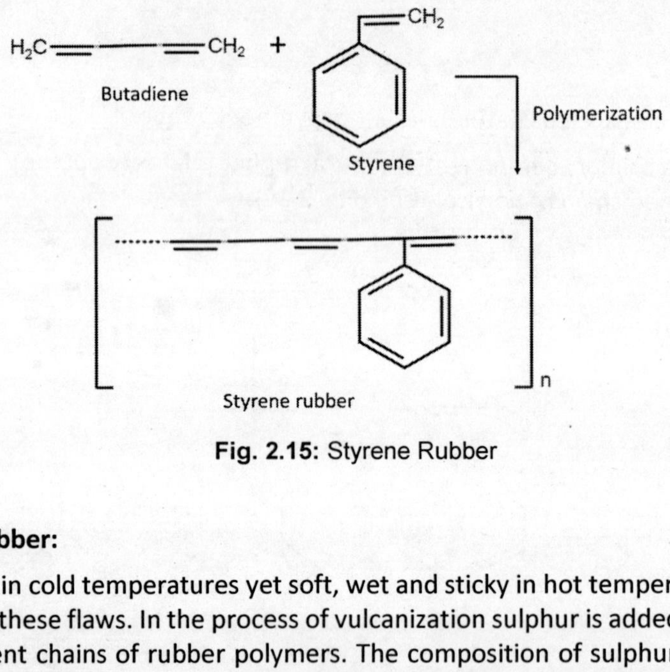

Fig. 2.15: Styrene Rubber

Compounding of Rubber:

Rubber is elastic in cold temperatures yet soft, wet and sticky in hot temperatures. Vulcanization is used to eliminate these flaws. In the process of vulcanization sulphur is added to create cross-links between the different chains of rubber polymers. The composition of sulphur varies based on the type of rubber being produced. For example, battery case rubber contains up to 30% sulphur, while vulcanised rubber tyres typically contain 3–5% sulphur.

The vulcanization process takes a long time, and the finished product is porous, weak, and degrades after being exposed to the air for an extended period. Before vulcanization, certain additives like vegetable oils, wax, and rosin oil are added to increase tenacity and adhesion. These additives are known as softeners. Other types of additives include accelerators, reinforcing pigments, and antioxidants. Accelerators like benzothiazole cut the time needed for vulcanization, antioxidants like complex amines prevent natural rubber from oxidising and degrading, and reinforcing fillers like carbon black provide strength and rigidity. Antioxidants or age resistors slow down deterioration and increase the rubber's lifespan. These are typically organic aromatic amine compounds that are added in very small amounts.

Uses of Rubber:

Rubber is used for a wide range of engineering applications, like conveyor belts, tyres and flooring tiles. The elastic properties of rubber make it an excellent shock absorber, which makes suitable for protective coatings and shock absorbing pads. Furthermore, rubber is also employed in the production of foam rubber, which is used in furniture and upholstery.

2.12 LAMINATE

Laminate is a composite material produced by stacking materials in different layers and bonding them using heat, pressure, welding, or adhesives. The process of making laminates is known as lamination. The different layers of laminates can be made from different materials, like decorative laminate board, or the same material, like plywood.

Decorative laminate board

Plywood laminate flooring

Fig. 2.16: Examples of laminate

(Source: https://en.wikipedia.org/wiki/Laminate; https://en.wikipedia.org/wiki/Laminate)

Different types of laminates:

Laminates can be differentiated on the basis of their applications, material used manufacturing process, and several other factors. Some of the commonly used forms of laminates are as follows:

Decorative laminates: Decorative laminates are identified by their intended application. These laminates are hard sheets, generally of 1 mm thickness, used to provide an aesthetic appearance to furniture elements. Decorative laminates can be designed to have many different colours, patterns, and textures. Decorative laminates are often used in residential areas to enhance the aesthetic appeal of the wooden furniture.

Fig. 2.17: Decorative laminate

(Source: https://en.wikipedia.org/wiki/Decorative-laminate)

Industrial laminates: Industrial laminates are identified by their intended application. Industrial laminates are typically made by embedding fibres in thermosetting resins, and then stacking them in multiple layers. The industrial laminates are more durable, stronger and low electrical conductivity. They are also scratch, wear, chemical, microbial, and fire resistant. As a result, industrial laminates are commonly used to make circuit boards, fixtures and thermal breaks.

High pressure laminate (HPL): They are identified by their manufacturing process. High pressure laminates are manufactured by bonding different layers at a high pressure, typically more than 5 MPa. During application of pressure, temperature may also be applied to improve the bond between the layers. High pressure laminates have exceptional durability and can support repeated loading, making them suitable for retail, hospitality, healthcare, and educational infrastructures.

Low pressure laminate (LPL): They are identified by their manufacturing process. Low pressure laminate is manufactured at low pressures, typically less than 3 MPa along with application of heat. Low pressure laminates generally involve soaking decorative paper in melamine resin before bonding laminated paper directly to particle board or fibreboard.

Compact laminate: They are identified by the stacking layers used to prepare the laminate. In compact laminate, a large number of layers are stacked in the middle to provide thickness and strength. The thickness of compact laminate sheets ranges from 3 to 30 mm whereas for regular laminate sheets this range varies from 0.8 mm to 1.5 mm. These laminates are self-supporting, thus wood or any other material does not need to be bonded to them.

2.13 ADHESIVES

Adhesives are a class of compounds which bind various materials and surfaces together, making it very difficult to separate them. They can be synthetic as well as natural. Natural binders, like plant juices, have been replaced by synthetic binders made from plant minerals or mineral oil, for most engineering applications. For structural applications, high load-bearing adhesives have been produced using rubber-toughened epoxies, acrylics, and silanes.

Advantages of adhesive:

One of the most important benefits of adhesive is its ability to effectively bond over different surface and create a joint. Other advantages include:

- Better aesthetics
- Low cost
- Reduced risk of corrosion in metal joints
- Ease of application

Disadvantages of adhesive:

At high temperatures, adhesives lose their stability and have poor peel resistance. They must be used with extreme caution in high temperature applications. Furthermore, adhesives take a long time to form a bond, and require special care until the adhesive sets.

Properties of Adhesive:

The adhesive properties of materials govern their applicability in adhesive processes. Some important properties of adhesives are as follows:

Adherence: To hold two surfaces together, the adhesive must form a strong bond at each interface between the surface and the adhesive, and the adhesive should also have high cohesive strength. Failure at either interface is possible because adhesive failure happens in the weakest bond. Failure can also occur within the adhesive, known as cohesive failure. In practice, bond failure at the interface is frequent. For the best results, clean surfaces are required regardless of the quality of the adhesive.

Wettability: Wettability defines the spread of adhesive on the applied surface. It must be able to flow into surface crevices and dispense with dirt, moisture, and trapped air. The surface tension of the adhesive can be reduced for this purpose by introducing a surfactant or wetting agent.

Fluidity: The viscosity and thixotropy of adhesives commonly define their fluidity. When adhesives are stirred or applied, the crystals rearrange, and the adhesives become more fluid. On the other hand, when the stirring is stopped, the fluid thickens. The ability to regain its stability in the absence of external action is defined as the thixotropy of adhesives. The thixotropy of adhesives help prevent sagging when these coatings are applied to vertical surfaces. Furthermore, the viscosity defines the ease with which material can flow and spread over any surface. Low viscosity helps in easier spread but does not stop the paint from running off when applying adhesives to vertical surfaces. Therefore, viscosity should be carefully maintained to support spread without causing runoff.

Types of adhesives:

Latex adhesives: Latex adhesives are natural or synthetic rubber or vinyl copolymers, containing elastomeric film formers, emulsifiers, thickeners, and antifoaming agents. Styrene-butadiene and neoprene are examples of synthetic rubber-based latex adhesives.

Organic solvent thinned adhesives: These adhesives use organic solvents as thinner to improve the fluidity and wettability of adhesives. They are typically applied to both surfaces, and joined after the solvent has evaporated. Ethyl acetate, methyl ketone and aliphatic paint thinner are some of the solvents used in adhesives.

Water based adhesives: These contains binder dispersed in water. Once the adhesives have been applied, the water dissipates and binding agent forms the joint. Different binders can be used for creating an emulsion in water, like, natural rubber adhesives and bio-based adhesives. Collagen, starch and casein are some examples of bio-based adhesives.

2.14 GRAPHENE

Graphene is a carbon allotrope, composed of a single layer of atoms structured in a two-dimensional honeycomb lattice. The name graphene is derived from *"graphite"*, where the suffix *"-ene"* indicates presence of double bonded carbon atoms. Graphene was first isolated in 2004, and since then graphene has evolved as a valuable and usable nanomaterial due to its remarkable electrical conductivity, tensile strength, and status as the world's thinnest two-dimensional material.

Fig. 2.18: Graphene

(Source: https://www.graphenea.com/pages/graphene#.YsVe-nZBxPY)

Properties of graphene:

Some notable properties of graphene are as follows:

- High hardness.
- High thermal and electrical conductivity.
- High density
- Antibacterial effect. Bacteria cannot thrive in it.
- High level of resistance.
- Graphene is around 200 times stronger than steel, with resistance equivalent to diamond but considerably lighter.

Application of graphene:

Graphene is extensively studied for engineering applications, due to the high availability of carbon in nature, and excellent properties shown by graphene. In civil engineering, several commercial applications of graphene are constantly being developed. For example, upgrading asphalt with graphene to create robust and thermally resistant roadways, developing cement grout materials that incorporate graphene for geothermal wells, and improvement in compressive, tensile, and flexural strengths of concrete by using surfactant functionalized graphene powder.

2.15 CARBON COMPOSITES

A carbon composite is an engineering material prepared embedding carbon fibres in a given matrix. Carbon composites in general offer low weight and significantly higher tensile strength as compared to their counterparts.

Organic polymers are used to create carbon fibre. These polymers are composed of continuous molecules-on-molecule chains that are joined by carbon atoms. Carbon fibres act as reinforcement for the matrix, which can be made from different materials. Fig. 2.19 shows the different carbon composites that are possible, on the basis of matrix being used.

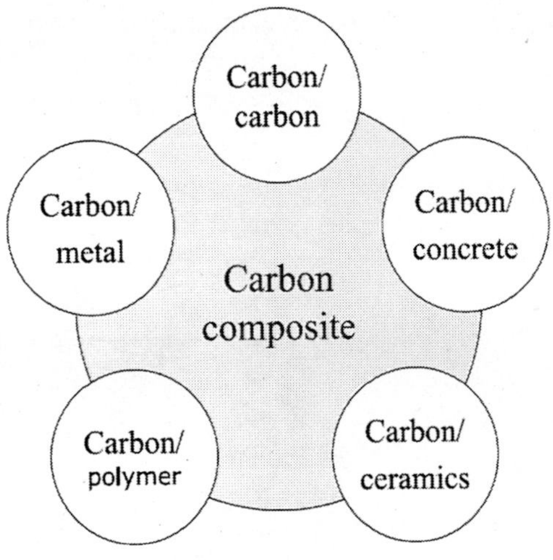

Fig. 2.19: Classification of Composites

Carbon/carbon composites:

Carbon fibres are encased in a carbon matrix to create a special type of composite material known as carbon/carbon composites. Carbon/carbon composites can have low density (1.6 to 2.0 g/cm^3) and high strength (more than 300 MPa) even at elevated temperatures (above 1400 °C). Furthermore, the strength of carbon/carbon composites increases with temperature, making them ideal for high temperature applications like building rockets and their launchpads. Carbon/carbon composites have been used for the development of Agni missile.

Properties of Carbon-Carbon Composites:

- Great strength at high temperatures
- High thermal conductivity
- Low frictional coefficient (in the direction of the fibre)
- Excellent resistance to thermal shock
- Low density
- High Modulus of Elasticity

Carbon composites have low oxidation resistance and reacts with oxygen at temperatures above 482 ºC. Hence, protective measures, like carbide coating, are performed for high temperature applications in presence of oxygen.

Carbon/concrete composite:

In carbon/concrete composite traditional metals like steel are replaced by carbon fibres as reinforcement. Carbon/concrete composite shows numerous advantages over conventional concrete, like less chance of rusting, light weight and better flexibility.

Carbon/concrete composites can be prepared in several different forms; some of which are as follows:

- Concrete is blended with carbon fibres, as in case of fibre reinforced concrete. The carbon fires are indiscriminately dispersed throughout the whole concrete matrix.
- Carbon fibre meshes, that resemble textile mesh, are incorporated between the thin concrete layers.
- The fibres can also be precisely shot into the concrete using the nozzle process. As a result, a material with an extremely high tensile strength is produced.
- Carbon fibre can be prepared as reinforcing bars and used in concrete.

Fig. 2.20: Concrete cube with carbon bar

(Source: http://www.bauen-neu-denken.de/presse/pressematerial/)

Carbon/ceramics composites:

Carbon/ceramic composites consists of carbon fibre embedded in ceramic matrices, typically made from silicon carbide. The primary objective is to provide robustness in ceramics and prevent their catastrophic failure. Carbon/ceramic composite material can withstand high operational temperature of about 1200°C. Carbon/ceramic composites have low density, corrosion resistance, and high-temperature oxidation resistance. Carbon ceramic composites have found a mainstream application of making brake discs for high-performance vehicles, as a substitute of cast iron.

Fig. 2.21: Carbon/ceramic composite brakes for racing car

(Source:https://www.cfccarbon.com/carbon-composite/carbon-ceramic-composite.html)

Carbon/polymer composite:

Carbon/polymer composites are made by embedding carbon fibre in polymer matrices, like plastic. These polymers have great resistance against corrosion. Carbon/polymer composites can be costly to produce, but they are frequently employed in areas where a high strength-to-weight ratio and stiffness (rigidity) are necessary.

Advantages of carbon/polymer composites include:

- Light weight
- Higher compression strength
- High corrosion resistance
- Resistance against extreme environmental conditions like chemical exposure, humidity and radiation.

The main disadvantage of carbon/polymer composite is its comparatively low fatigue life. However, recent studies have shown that modification in matrix can significantly improve the fatigue performance.

Carbon/polymer composites have several applications in different fields; some of them are as follows:

- As cables in suspension cable bridges and suspension dome system.
- In sporting goods for making rackets for tennis and badminton, skis, fishing rods, etc.

- In the aerospace industry, satellite bodies, aircraft structures, and wings are generally made of carbon/polymer composites.
- Several parts of the automobile are made up of carbon/polymer composites.
- In industries for making several medical instruments, robots, rotary blades and many more.

Carbon/metal composites:

It is a composite substance made of carbon fibres embedded in a metal matrix. Carbon/metal composites have better mechanical performance, thermal and electrical properties, resistance to corrosion and reduced coefficient of friction as compared to unreinforced metals. As carbon/metal composites have higher stiffness and strength, thinner and light weight sections can be used in load bearing applications.

Properties of carbon/metal composites are:

- Better mechanical strength
- Excellent electrical properties
- High thermal conductivity
- High corrosion resistance
- Low coefficient of thermal expansion
- Reduced coefficient of friction
- Improved wear resistance

Carbon/metal composites have great potential in aerospace, automobile and petrochemical industries. In civil engineering, they can be used for improving the performance of steel (metal) structures and elements.

UNIT SUMMARY

Introduction to various emerging civil engineering materials

- Ceramic, Refractories, Plastic, Aluminium, Copper, Tin, Nickel, Acoustical material, Geotextile, Rubber, Laminate, Adhesives, Graphene and Carbon composites.
- Different variations of each engineering material.

Process of developing engineering materials

- Raw materials used and their properties.
- Different production techniques.
- Possible modifications and treatments with their effect on the final engineering material.

Properties of engineering materials

- Strength based – compressive, tensile, flexural, shear, split tensile, fatigue, impact, ductility
- Others - weight, density, colour, opacity, flowability, setting, hardening, thermal conductivity, acoustic, durability
- Important properties for different engineering materials

Application of engineering materials

EXERCISES

Multiple Choice Questions

2.1 Which of the following is false for ceramics

(a) Good conductor of heat and electricity (b) Refractory in nature

(c) Corrosion resistant (d) Brittle material

2.2 Nickel is which type of metal

(a) Brittle (b) Ductile (c) Semi-brittle (d) None of the above

2.3 Sound intensity for acoustic materials is measured in

(a) cm (b) kg (c) dB (d) kN

2.4 Coagulation of rubber in presence of retarder, such as sodium bisulphite, results in the formation of

(a) Crepe rubber (b) Smoked rubber

(c) Synthetic rubber (d) Latex

2.5 The primary component of natural rubber is

(a) Butadiene (b) asbestos (c) latex (d) None of the above

2.6 Which of the following act as a permeable cloth during drainage from road pavements

(a) latex (b) asbestos (c) geotextiles (d) rubber

2.7 What is the disadvantage of carbon/polymer composite

(a) Low strength to weight ratio (b) Low resistance to corrosion

(c) Low strength (d) Low fatigue endurance limit

2.8 Which of the following alloys is not made from copper:

(a) Gun metal (b) Aldural

(c) Brass (d) Bronze

2.9 Which of the following material is excellent in fire resistance

(a) Geotextile (b) Rubber

(c) Aluminium (d) Refractories

2.10 Which of the following is false for graphene

(a) Graphene is around 200 times stronger than steel

(b) Graphene is made of carbon and hence susceptible to bacterial attack

(c) Graphene is a sheet like material made of carbon allotrope

(d) Graphene has high thermal conductivity

Answers of Multiple Choice Questions

Answers of Multiple Choice Questions
2.1 (a), 2.2 (a), 2.3 (c), 2.4 (a), 2.5 (c), 2.6 (c), 2.7 (d), 2.8 (b), 2.9 (d), 2.10 (b)

Short and Long Answer Type Questions

2.1 Classify plastics based on the manufacturing process.

2.2 Write a short note of different alloys of aluminium.

2.3 Compare sound proofing and sound absorption.

2.4 Describe the various functions of geotextile.

2.5 List the engineering materials which can be used for elevated temperature, when exposure temperature crosses (a) 300 °C and (b) 800 °C.

2.6 Write short note on rubber.

Numerical Problems

2.1 An acoustic material provides noise intensity reduction in proportion of its thickness, given by the following formula;

$$\text{reduction in sound energy level (\%)} = 4.6^{-\,thickness\ (in\ mm)}$$

Calculate the thickness of material required to be placed for a library (≤ 30 dB) near an active traffic site (100 dB).

KNOW MORE

Smart materials

Among various emerging civil engineering materials there is a category of materials known as smart materials. Smart materials are a class of material which can sense the change in environment and respond according to it. Examples of smart materials include colour changing paints, tinting glass and self-healing concrete. These are some of the smart materials which are already being used in engineering applications.

Smart materials are adaptive and versatile which makes their application highly advantageous for civil engineering applications with changing stimulus. For example, a self-healing concrete is able to sense cracks in concrete and repair them. The typical application of a smart material is governed by the nature of stimulus being exposed to the material. The different types of stimuli affecting smart materials are strains, temperature, moisture, chemical composition, electrical field, magnetic field and sunlight.

One should keep in mind that smart materials can be made by both with and without electronic components. An awareness of this materials will help in their use and significantly improve engineering applications.

REFERENCES AND SUGGESTED READINGS

1. V. Raghavan, *Materials Science and Engineering*, PHI Learning Pvt. Ltd., 2021.
2. I. P. Singh, S. Chander and R. K. Prasad, *Materials Science and Engineering*, Jain Brothers, 2020.
3. R. K. Sharma, *Testing of Construction Materials*, Wiley, 2019.
4. J. P. Patel and G. H. Upadhyay, *Material Science*, Atul Prakashan, 2001.
5. M. L. Gambhir and N. Jamwal, *Building and Construction Materials* , McGraw Hill education, 2014.
6. S. K. Duggal, Building Materials, New Age International (P) Limited, Publishers, 2019.
7. N. Subramanian, *Building Materials Testing and Sustainability*, Oxford University Press, 2019.

Dynamic QR Code for Further Reading

3 Introduction to Material Testing

UNIT SPECIFICS

Through this unit, we have discussed the following aspects:

- *Various mechanical behaviours and characteristics of materials;*
- *Different types of strength of materials*
- *Fracture mechanics*
- *Various types of deformation, including modes of failure;*
- *Principles behind material testing*

The different types of mechanical behaviour and characteristics of civil engineering materials are discussed to create an understanding for the material testing (strength, deformation and failure modes) and encourage their safe application.

Along with a large number of multiple-choice questions and questions with short and long answers, the unit also includes a list of references and suggested readings that one can use as practice and grow their knowledge on material testing.

RATIONALE

This fundamental unit on material testing helps the reader to familiarize with the different behaviours and characteristics exhibited by civil engineering materials. The unit describes the various strength, deformations and associated characteristics of the material such as tensile strength, creep, proof strain and elasticity, along with the introduction to fracture mechanics. The unit describes the concepts of material testing, important mechanical properties and their interpretations. The discussion is followed by some problems related to material testing, which will further help in developing a clear idea of the concerned topics on the subject.

A good grasp of materials and their behaviour is necessary for the construction of buildings and other structures. Understanding the mechanical behaviour of different materials help in safe engineering application of the said material. A better understanding of material testing will help engineers in understanding the necessary mechanical behaviour of different materials.

PRE-REQUISITES

Physics: Mechanics (Class XII).

Chemistry: Lattice structures (Class XII).

UNIT OUTCOMES

As outcome of this unit, one should be able to:

U3-O1: Understand mechanical characteristics and behaviour of engineering materials.

U3-O2: Appreciate the concepts of elastic and plastic deformations.

U3-O3: Study the deformation and failure pattern under different loading patterns.

U3-O4: Comprehend the concepts of fracture mechanics.

U3-O5: Apply appropriate material testing for desired engineering purposes.

Unit-3 Outcomes	EXPECTED MAPPING WITH COURSE OUTCOMES (1- Weak Correlation; 2- Medium correlation; 3- Strong Correlation)							
	CO-1	CO-2	CO-3	CO-4	CO-5	CO-6	CO-7	CO-8
U3-O1	-	1	-	2	2	2	2	2
U3-O2	2	2	-	1	3	3	2	2
U3-O3	2	2	3	2	2	2	3	3
U3-O4	1	-	2	1	-	-	3	2
U3-O5	3	2	3	2	-	-	-	2

3.1 MATERIAL ENGINEERING

Material engineering is a field involving material properties and their applications to numerous fields of engineering and research. This includes fields like civil engineering, electrical engineering, chemical engineering, mechanical engineering, construction, aeronautical, agricultural and biomedical. Any engineered product, structure, or device need the appropriate materials to be formed. Material engineering helps in understanding what things are made up of and what are the reasons for that behaviour. Material engineering teaches us how to use knowledge of an existing material to improve them in order to get better performance. This field of study focuses on the relation between a materials composition (including its atomic or molecular structure) and its macroscopic characteristics. Understanding the composition of material, their optimal uses, and how to combine them for greater performance is made feasible by material engineering. For a material, relation between desired qualities and relative performance in a particular application forms base of material engineering.

3.2 MECHANICAL BEHAVIOR AND MECHANICAL CHARACTERISTICS

A material's deformation in relation to externally applied loads or forces is reflected in its mechanical behaviour. Strength, hardness, stiffness, and ductility are one of the most important mechanical characteristics. The properties that affect a material's response to applied loads are referred to as its mechanical characteristics. Mechanical characteristics are used to predict how a material will behave in a specific application and are useful when choosing a material.

Following are various mechanical characteristics:

Strength: Strength is a material's capacity to withstand deformation. Withstanding deformation can imply both whether the material can withstand a load without changing shape or whether it can do so without breaking. Materials that we choose for our engineering solutions need to be sufficiently strong to function under a variety of mechanical forces or loads.

Elasticity: Elasticity is defined as a property of any material that allows it to restore its original size and shape when a load has been entirely removed. When an external force is applied on a body, it undergoes some deformation. If the body returns back to its original shape and size and position after the complete removal of an external load, then the body is called as an elastic body. The majority of solid materials display some degree of elastic behaviour, but for each material there is a maximum force and deformation range within which elastic recovery is achievable.

Toughness: The capacity of any material to absorb energy and undergo plastic deformation without fracture or failure is known as toughness. A stress-strain diagram can be used to define toughness in terms of areas. The area under the stress-strain curves up to the point of failure is known as toughness. Materials should have good strength and ductility in order to be tough.

Malleability: The ability of a material to get rolled or hammered into a thin sheet is called as malleability. This quality is a component of material's plasticity. This property is mainly affected by the temperature. Malleability of any material increases with rise in a temperature.

Ductility: Ability of a material that allows it to be pulled into a thin wire is known as ductility. This mechanical characteristic, which is temperature-dependent, is also an important aspect of a plasticity of any material. ductility increases with rise in a temperature.

Creep: Creep is a permanent deformation that occurs with time due to constant loading or pressure. It is mostly caused by prolonged exposure to relatively higher stresses that are within the yield limit and is more severe in materials that have been heated over an extensive period. When the stress magnitude is getting close to the fracture failure, creep becomes a considerable problem. If rate of creep rate increases rapidly, the strain could become high such that it results in failure.

Fatigue: The term "fatigue" refers to the material's weakening mainly caused by the repetitive loading and unloading. Micro cracks start to form at grain boundaries and interfaces when a material is subjected to cyclic type loading that is above a threshold magnitude but significantly lesser than the material's ultimate tensile strength. The crack eventually propagates to be too large. This propagation takes place quickly and it results in a fracturing of the material. The structure's form significantly affects the fatigue.

Resilience: A material's capacity to absorb energy without getting any permanent deformation is known as resilience. When the body is within its elastic limit, the strain energy that is stored in the body as a result of an external load is called to as resilience. For applications where materials must be flexible under stress yet regain their original shape once the force is removed, resilience is a significant material selection criterion for a large number of materials.

Plasticity: The plasticity of a material is defined as its ability to undergo permanent deformation without causing rupture or brittle failure. Plastic deformation will occur only after the elastic range has been exceeded.

Hardness: A material's ability to withstand scratching, abrasion, cutting, or penetration is referred to as hardness. From an engineering perspective, hardness is one of the important characteristics because resistance offered to wearing action caused by friction or erosion induced due to water, oils and steam increases as the hardness increases

Brittleness: A material's lack of ability to withstand permanent deformation is defined as its brittleness. Brittle materials do not undergo plastic deformation and directly fail under breaking. Two examples of brittle materials are concrete and glass. Higher compressive strength, lower tensile strength, low deformation, poor resistance to impact and vibration of load are the characteristics of brittle materials.

3.3 ELASTICITY – PRINCIPLE AND CHARACTERISTICS

Elasticity is defined as a property of any material that allows it to regain its original size and shape when a load has been entirely removed. The elastic limit of a material is nothing but a maximum stress that can be applied to it before it is permanently deformed and does not return to its initial size and shape. For the greater advancement of numerous engineering and scientific fields, such as mechanical, civil, electronic, aeronautical, aerospace, and others, it is crucial to study the elastic properties of diverse materials and look for new materials.

One needs to derive a number of boundary conditions and fundamental equations in order to analyse the displacements, stresses, strains which are related to elasticity problem. Although one can take into account all the influencing aspects when formulating such equations, the results would still be extremely complex, making it nearly impossible to find a solution. Therefore, in order to arrive at potential solutions, some of the assumptions on a fundamental basis regarding the characteristics of the body under consideration must be made. With such assumptions, we can neglect some of the important but less significant secondary influences.

The following are the assumptions in elasticity:

- The body is assumed to be perfectly elastic: The body is assumed to obey Hooke's law of elasticity, which represents the linearity between the stress and strain. as per this assumption, the elastic constants are considered not to be dependent on the magnitudes of stress and strain components.
- The body is assumed to be homogenous: All parts of body have similar elastic characteristics. As a result, the location in the body will not affect the elastic constants. According to this principle, one can study a body component that has been separated before applying the findings to the complete body.
- The Body is assumed to be Isotropic: A body's elastic characteristics are constant in all direction. As a result, the coordinate axes orientation will have no effect on the elastic constants.
- The displacements and strains are small: When a body deforms, all of its point's displacement components are very small compared to their initial sizes, and all of the line elements' rotation components are substantially smaller than one.

3.3.1 Linear and nonlinear theory of elasticity

The nonlinear theory of elasticity corresponds to the fact that the relationship between stress and deformation is nonlinear. The material is described as linearly elastic if the relation between stress and deformation is linear, and the corresponding theory is known as the linear theory of elasticity.

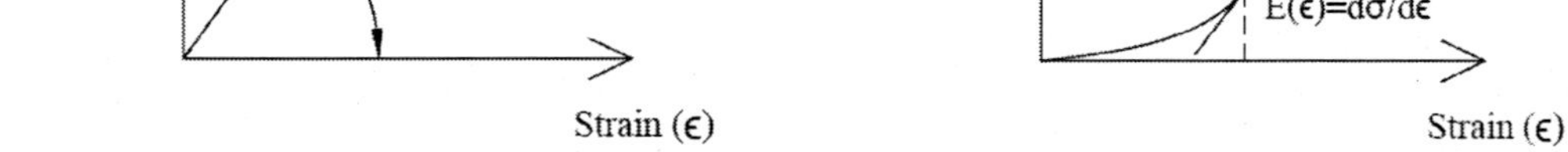

(a) linear elasticity (b) nonlinear elasticity

Fig. 3.1: Linear and non-linear elasticity

3.3.2 Stress

Stress is the measure of material's response towards an externally applied force. It is defined as a ratio of the externally applied loads and the cross-sectional area of the material on which it is acting. its unit is $\mathrm{kN/m^2}$

Stress can be calculated by,

$$\text{stress } (\sigma) = \frac{\text{Force } (F)}{\text{Area } (A)} \tag{3.1}$$

Types of stress:

Tensile stress: Tensile stress is a state of tension brought on by an applied force that attempts to elongate the material along its axis.

Compressive stress : Compressive stress (also known as compression) is a state of stress brought on by an applied force that shortens the length of the material along its axis. The uniaxial compression brought on by the action of opposing pushing forces is a simple example of compression.

Shear stress : Shear stress is the stress condition caused by the combined force of two opposing forces acting through the material in parallel directions. It is the strain brought on by the material faces sliding against one another.

Bending stresses: The resistance offered by an object to the action of bending due to load is known as the bending stresses. When an object bends, a normal load is delivered at a particular location on the object. In other terms, a bending moment is a measure of the internal bending effect caused by external loads.

Torsional stress: Shear stress brought on by twisting is known as torsional stress. It could also be referred to as a body's angular distortion.

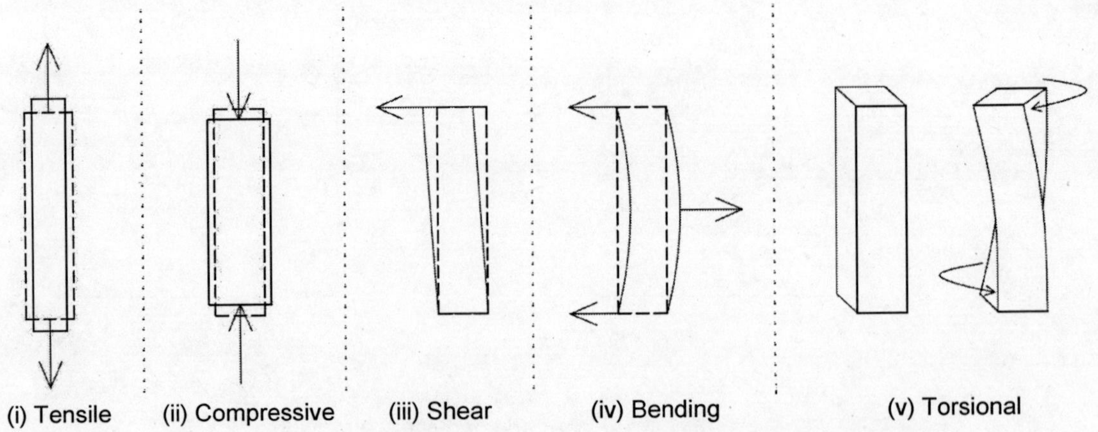

Fig. 3.2: Types of loadings that causes different types of stresses

3.3.3 Strain

A strain is defined as a ratio of change in dimension to the original dimension. Strain is a unitless quantity, it is the parameter that defines the intensity of internal deformation. It can be further classified into normal strain and shear strain.

It can be formulated as,

$$\text{strain } (\epsilon) = \frac{\text{change in dimension}}{\text{original dimension}} = \frac{\Delta l}{l} \tag{3.2}$$

Types of strain:

Tensile strain: Tensile strain is the term used to describe the strain that a tensile force causes in a body. The body's cross-section area decreases and its length always increases in response to tensile tension. Tensile strain in this case is defined as the proportion of the increased length to the initial length.

Compressive strain: Compressive strain is the strain that develops as a result of a compressive force. The dimension of the body shrinks under compressive force. Compressive strain is defined as the difference between the body's reduced length and its initial length.

Volumetric strain: Volumetric strain is the ratio of a body's changed volume to its initial volume. In volumetric strain, the body's volume changes as a result of the application of external forces.

Shear strain: The strain which is caused in a body due to the application shear force is called shear strain. A change in angle between lines that were once perpendicular is how shear strain is measured.

3.3.4 Concepts associated with elastic deformation

Hook's law: According to Hooke's law, within a limit of proportionality stress is directly proportional to corresponding strain, i.e., stress $\propto$ strain.

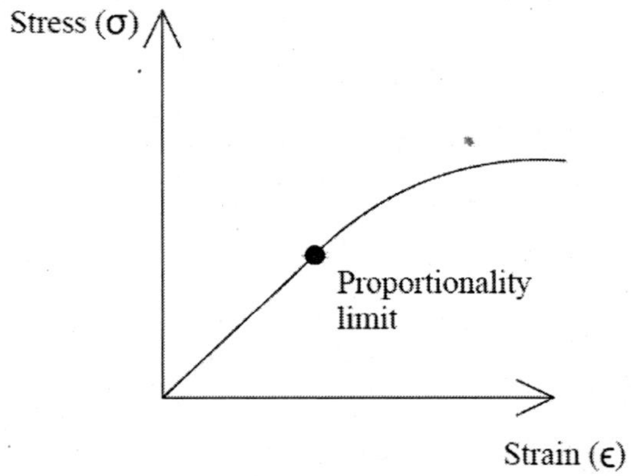

Fig. 3.3: Limit of proportionality in stress vs. strain graph

Modulus of elasticity or Young's modulus: The ratio of applied stress to the strain is known as modulus of elasticity or Young's modulus and it is denoted by letter "***E***". The unit of modulus of elasticity is $\mathbf{kN/m^2}$, which is same as that of stress.

$$\text{Modulus of elasticity } (E) = \frac{\text{stress } (\sigma)}{\text{strain } (\epsilon)} \tag{3.3}$$

Bulk modulus: When a body is subjected to mutually perpendicular direct stresses which are alike and equal, within its elastic limits, the ratio of direct stress to its corresponding volumetric strain is found to be constant. This ratio is known as bulk modulus and is represented by letter "***K***". Its unit is MPa.

$$\text{Bulk modulus}(K) = \frac{\text{direct stress}}{\text{volumetric strain}} \tag{3.4}$$

Rigidity modulus: When a body is subjected to shear stress the shape of the body gets changed, the ratio of shear stress to the shear strain is known as shear modulus or modulus of rigidity. It can be represented by the letter's "***G***" or "***C***". Its unit is MPa.

$$\text{Rigidity modulus } (G) = \frac{\text{shear stress}}{\text{shear strain}} \tag{3.5}$$

Poisson's ratio: The Poisson's ratio is defined as a ratio of lateral strain to longitudinal strain. It is represented by the symbol "$\boldsymbol{\nu}$". For a perfect elastic incompressible material, Poisson's ratio is maximum and has a value of 0.5. The Poisson's ratio for the majority of engineering materials ranges from 0.25 to 0.33. It is a unitless quantity.

$$\text{Poisson's ratio } (\nu) = \frac{\text{lateral strain}}{\text{longitudinal strain}} \tag{3.6}$$

Representation of relationship between elastic constants

- The relationship between Poisson's ratio (ν), rigidity modulus (G) and Young's modulus (E) can be expressed in following form:

$$E = 2G\,(1 + \nu) \tag{3.7}$$

- The relationship between bulk modulus (K) and Poisson's ratio (ν) and Young's modulus (E), is can be expressed in following form:

$$E = 3K\,(1 - 2\nu) \tag{3.8}$$

- Young's modulus (E) can be expressed in terms of rigidity modulus (G) and bulk modulus (K) in the following form:

$$E = \frac{9KG}{(3K + G)} \tag{3.9}$$

- Poisson's ratio (ν) can be represented in terms of rigidity modulus (G) and bulk modulus (K) as:

$$\nu = \frac{(3K - 2G)}{(6K + 2G)} \tag{3.10}$$

3.4 PLASTIC DEFORMATION OF METALS

A body that has its elastic limit exceeded will either permanently deform or set when the applied load is removed. When an object is permanently deformed, it is referred as plastic deformation. It occurs when atom's bonds break and new one's form, making it hard for the atoms to return to their previous structure. Plastic deformation produces the product shapes in metal forming processes. In order to optimize the processes, it is crucial to understand the plastic flow characteristics of metals and alloys. Additionally, the degree and circumstances of plastic deformation during the forming process affect the properties of the final component.

Mechanism of plastic deformation: Plastic deformation of metals is basically studied under two categories namely,

1. Plastic deformation of single crystals.
2. Plastic deformation of polycrystalline materials

3.4.1 Plastic deformation of single crystals

The study of one particular crystal and observation of how it responds to stress under plastic deformation of single crystals. a single crystal is consisting of just a single grain with no grain boundaries. In single crystals plastic deformation takes place by;

1. Slip
2. Twinning

Plastic deformation by slip

Slip is also known as dislocation motion. Among crystal it is the most common type of plastic deformation. A step that causes the single crystal to split into two blocks arises when a single crystal in tension is pushed past its elastic limit. When the tensile load is increased further, the blocks split once more, and relative displacement occurs. Slip always occurs on a particular set of crystallographic planes, known as slip planes. These directions, which are known as the slip directions, are the same for all slide inside these planes. A slip system is composed of the components slip plane and slip direction.

A specific set of crystallographic planes known as the slip planes, is where the slip usually takes place. These directions, which are known as the slip directions, are the same for all slide inside these planes. A slip system is composed of the components slip plane and slip direction.

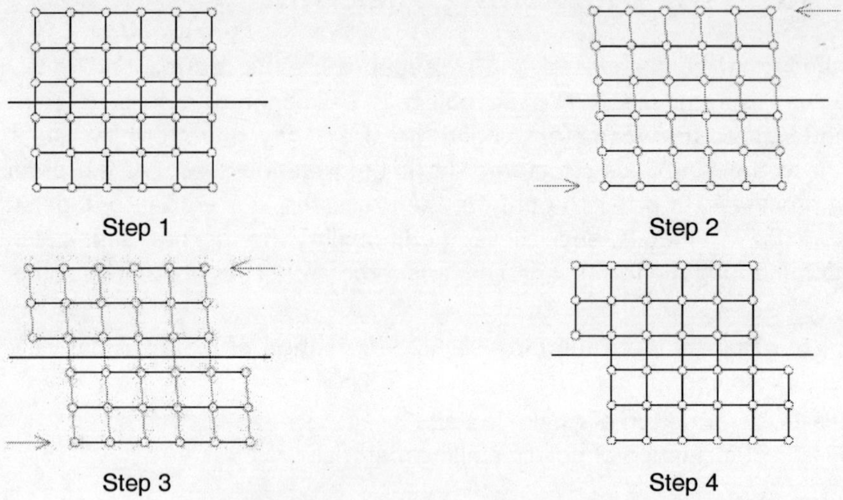

Fig. 3.4: Slip process in a single crystal

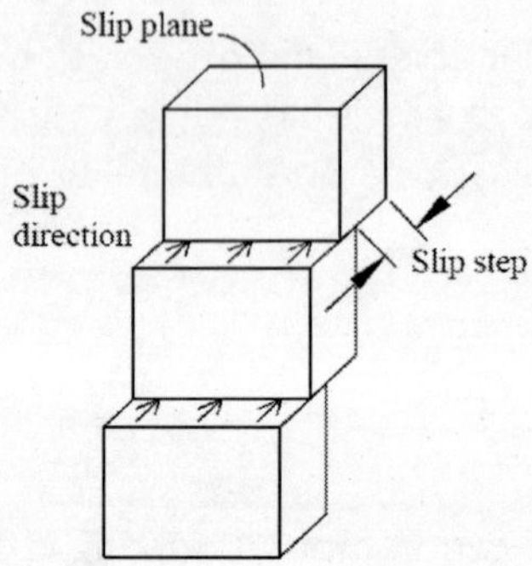

Fig. 3.5: Deformation by slip

Slip system:

The family of slip directions and set of symmetrically similar slip planes where dislocation motion is easily generated and results in plastic deformation. It is the main criterion for a material's plastic deformation, which could increase the chances of failure.

On the basis of lattice structure of the crystals, there are three types of slip systems:

1. **Face Centred Cubic (FCC) slip system:** In FCC crystals, the lip occurs on the close packed planes, i.e., the plane for maximum atomic density. There are four possible planes of slip in FCC namely, $\{1\ \ 1\ \ 1\}$, $\{1\ \ \bar{1}\ \ 1\}$, $\{\bar{1}\ \ 1\ \ 1\}$ and $\{1\ \ 1\ \ \bar{1}\}$. Along any given plane, the slip can occur in 6 possible directions, i.e., $\langle\bar{1}\ \ 1\ \ 0\rangle$, $\langle\bar{1}\ \ 0\ \ 1\rangle$, $\langle 1\ \ \bar{1}\ \ 0\rangle$, $\langle 0\ \ \bar{1}\ \ 1\rangle$, $\langle 1\ \ 0\ \ \bar{1}\rangle$ and $\langle 0\ \ 1\ \ \bar{1}\rangle$. While total slip directions appear to be 24, each slip direction is common for two planes of slip. Therefore, in FCC crystals, there are 12 slip systems. Because there are so many slip systems, it is easy to deform FCC crystals because there is always some plane or direction that is orientated in a way that allows plastic deformation.

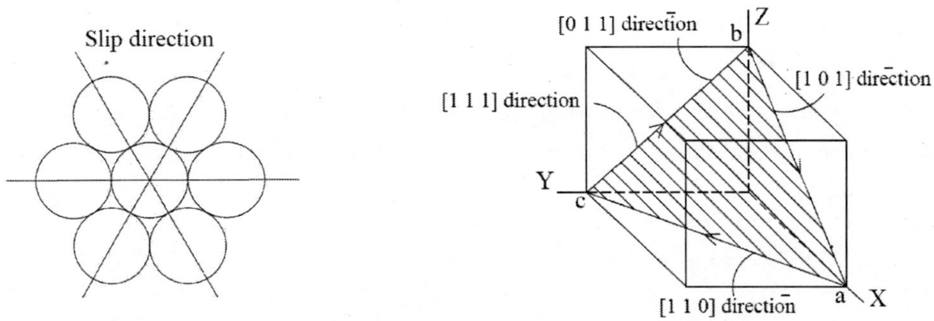

(a) Easy slip direction (b) Slip system

Fig. 3.6: Slip system in FCC crystals

2. **Body Centred Cubic (BCC) slip system:** In BCC crystals, slip predominantly occurs along the 6 $\{1\ \ 1\ \ 0\}$ planes in 2 possible $\langle 1\ \ 1\ \ 1\rangle$ directions. Similar to the 12 $\{1\ \ 1\ \ 0\}$ slip systems, there is a possibility that sip can also occur in 12 $\{2\ \ 1\ \ 1\}$ or 24 $\{3\ \ 2\ \ 1\}$ slip systems.

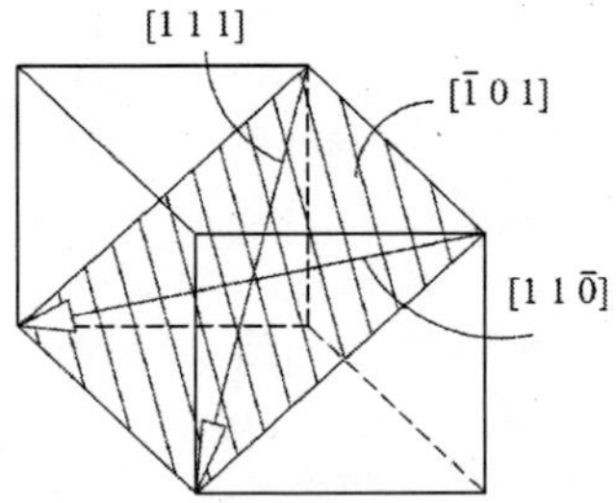

Fig. 3.7: Slip system in BCC crystals

3. **Hexagonal Close Packed (HCP) slip systems:** In HCP crystals, there is only one plane having high atomic density for slip, known as the basal plane $\{\mathbf{0\ 0\ 0\ 1}\}$. The three axes $<\mathbf{1\ 1\ 2\ 0}>$ or close-packed directions lying on this plane along which slip occurs. Thus, HCP crystals have 3 slip systems. The slip in HCP crystals is less as compared to FCC metals.

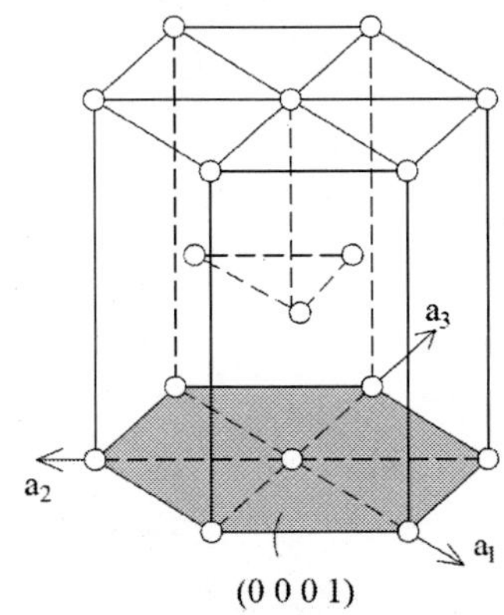

Fig. 3.8: Slip system in HCP crystals

Schmid's laws (critical resolved shear stress):

Let us consider the cylindrical single crystal which is having area A and P be the axial force which will act upon it. consider the λ as an angle between the slip direction and and tensile axis i.e stress direction and Φ be the angle between normal to slip plane and tensile axis. $P \times \cos(\lambda)$ will be the component of axial force that acts in the slip direction on the slip plane. This component of force is effective in moving the dislocation. The area of slip plane is $A/\cos(\Phi)$, and the critical resolved shear stress τ_c is given by,

$$\tau_c = \frac{P \times \cos(\lambda)}{A/\cos(\Phi)}$$

$$\text{or} \quad \tau_c = \sigma \times \cos(\lambda) \times \cos(\Phi) \tag{3.11}$$

As the temperature increases the magnitude of τ_c decreases. The stress required to initiate slip at a given temperature in a pure and perfect single crystal is constant for a material, this is known as the Schmidt's law.

Plastic deformation by twinning

Each plane of atoms moves through twinning over a specific distance and in the same direction. Each plane moves in a manner that is proportionate to how far away it is from the other plane. The crystal will twin about the twinning plane under a shear force such that the area to the left of the twinning plane is not distorted while the area to the right is. The atomic configuration is such that the twinned plane's two sides are mirror images of one another

The twinned section, participate in deformation by three ways:

1. Twinning helps in changing the shape of service
2. Reorientation of the crystal region between the twin planes.
3. By placing the potential slip planes into better position, twinning helps in further slip.

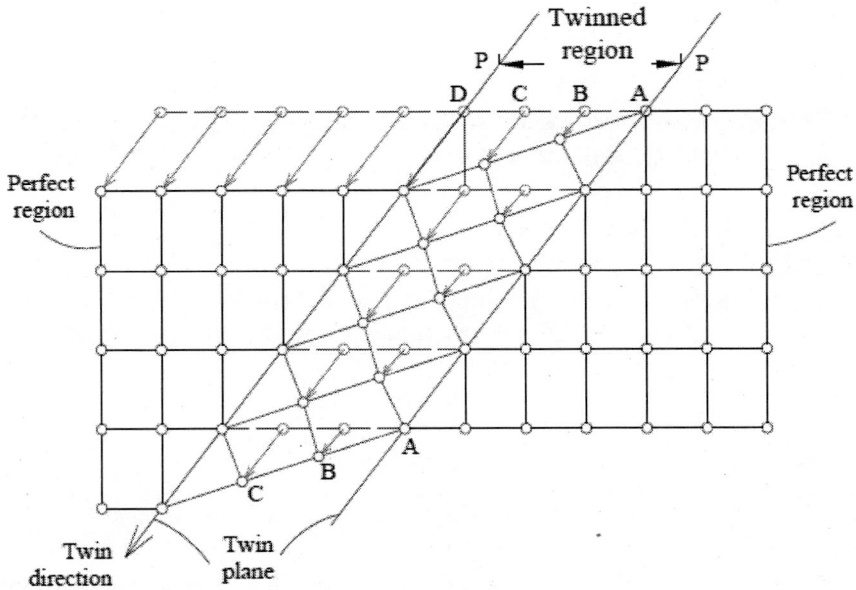

Fig. 3.9: Twinning in crystal

3.4.2 Plastic deformation of polycrystalline materials

As the effects of grain borders and neighbouring crystals are taken into consideration, deformation in polycrystalline material is more complicated than in single crystal. When crystal is of geometric regularity, the same slip system operates. Dislocations are more mobile when there is a divergence from the ideal crystal structure. The arrangement of atoms in a region of a grain border differs greatly from that found in a single perfect crystal because the grain boundary is a surface at which two crystals of different orientation meet. The impact of the grain boundary on the dislocation's positions is to immobilise them by applying a repulsive force to each succeeding dislocation that descends the slip plane. This results in dislocation accumulation due to the grain boundary. The electron microscope can be used to observe such pile-ups. Polycrystals are often stronger and less ductile than single crystals because the grain boundaries reduce dislocation mobility. Changes in the grain boundaries increases the resistance to static fracture, fatigue, and creep. Some of the crystals in polycrystalline materials behave as single crystals because their slip directions are aligned in the plane of maximum shear. As a result, these crystals exhibit slight yielding when compared to their more elastic neighbours. In this type of polycrystalline materials, initial yielding is relatively gradual. Grain size and orientation is another factor influencing the mechanical properties in early stage of deformation in polycrystalline material. In polycrystalline materials, deformation is never uniform and the reason for that is random orientation of the grains. So, the slip system gets affected because of this. When the grains are smaller, there are more grain boundaries for dislocations to flow through, and there will be more barriers for dislocation motion. Hence yield strength is more dependent on grain size than tensile strength in early stages of deformation. However, for latter stages of deformation, it is the interaction of dislocations of deformation within the grains on which the strength highly depends.

Following are the characteristics of polycrystalline metals:

- Under the state of loading various grains shows the random orientation for the slip
- Due to the random orientation greater stress is required, hence the resolved shear stress is not constant but varies with the orientation of the grain.
- Within the crystals there are number of slip planes.
- All crystals are perfectly elastic under a load.
- Deformation results in a considerable structural change.
- Elastic after effect is observed in these metals which cannot be obtained in single crystal.
- Significant amount of permanent deformation and considerable amount of yielding can be obtained in the crystals.

3.5 STRENGTH

Strength of the material is defined as its ability to resist the applied load or stress without failure. Failure of a material can be defined either as breaking of material due to excess loading or as excess deformation in the material. Depending on the required application, various different types of strengths are used for any given material. The different types of strengths are typically classified by the nature of stress defining failure and characteristic strain defining the said failure. In terms of nature of stress, the strength can be tensile, compressive, flexural, etc. In terms of characteristic strain defining the failure strength is defined as ultimate, yield, etc. In order to determine the strength, the material is subjected to various degrees of strain and corresponding stress is observed. The stress-strain curve is then used for determining the strength of materials.

3.5.1 Tensile strength

Tensile strength is determined using a destructive mechanical testing procedure, known as tensile test. In this test, the resistive force against applied tensile load is measured, until a point where a material cannot elongate further by its inherent strength. Tensile test is important such that it provides relevant information for selection of materials, evaluation of their tensile properties and for quality assurance purposes. Tensile test is performed to obtain various properties like the tensile strength, yield strength, elongation (in percentage), change in area (in percentage), failure load, modulus of elasticity and maximum load at failure.

The main tensile testing standards for different materials are listed below:

- **IS 1608**: Mechanical testing of metals – Tensile Testing
- **ASTM E8/E8M**: Tensile Testing of Metallic Materials
- **BS EN ISO 4136**: Destructive tests on welds in metallic materials. Transverse tensile test
- **BS EN 895**: Destructive tests on welds in metallic materials. Transverse tensile test
- **ISO 6892**: Tensile Testing of Metallic Materials (IS 1608: 2005)
- **ASTM D412**: Tensile Testing of Elastomers
- **ISO 37**: Tensile Testing of Elastomers
- **ASTM D638**: Tensile Testing of Plastics

- **ISO 527-2**: Tensile Testing of Plastics
- **ASTM A370**: Standard Test Methods and Definitions for Mechanical Testing of Steel Products
- **ISO 377**: Steel and steel products – preparation of samples and test pieces

Tensile tests apparatus

In this test, the ends of the test piece of standard dimensions are fixed into grips connected to a strain-gauge device, and also to a load measuring device. The strain-gauge of this Universal testing machine (UTM) as shown in Figure comprises of a hydraulic cylinder with a robust base and an up-down movable piston. The lower table is connected to the piston through a ball, along with a ball seat to ensure axial loading. The lower and upper tables move up and down with the piston. The loading-gauge consists of a large dial and a weightless pointer to determine maximum loading reached during the test. Load is then applied through a hydro-static ram, at a constant rate of strain. The cylinder pressure is transferred to cylinder of pendulum dynamometer system, engaged with the control panel, which consists of a drive motor as well as an oil tanker, control valves and an automatic strain recorder. The pressure from the dynamometer is then transferred to the pendulum which initiates the racking and pinion mechanism, much similar to the steering mechanism of wheeled vehicles, thus the strain gauge and the load pointer also actuates. Pendulum deflection represents that loading have been applied to the specimen. The return movement of the pendulum is damped effectively to absorb energy of the pendulum, in case any emergency breakage of specimen occurs.

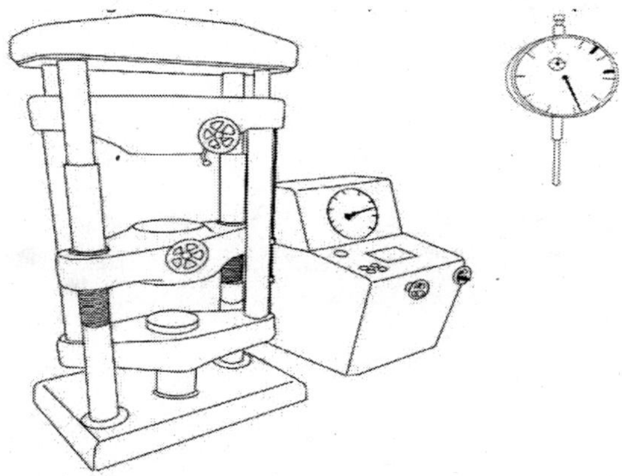

Fig. 3.10: Universal testing machine for tensile tests
(https://sm-nitk.vlabs.ac.in/exp/tensile-test-mild-steel/procedure.html)

Brittle materials:

Materials which have post-elastic strain less than 5% are said to exhibit brittleness. These materials don't have any definite yield point, and don't have any plastic region. Hence. The ultimate strength and fracture strength of such materials are same. Hence, these materials are also unable to absorb any energy before breaking.

Fig. 3.11 shows the graphical representation of typical cast iron specimen, which is brittle in nature. This Figure illustrates that brittle materials do not yield. Hence, their yield stress is determined using offset method. In this method, an imaginary line is drawn at a prescribed strain of 0.002 (0.2%), parallel to the tangent of the initial portion of the stress-strain curve. The point of intersection of this parallel line and the graph obtained by tensile testing gives us the proof strength/ yield point.

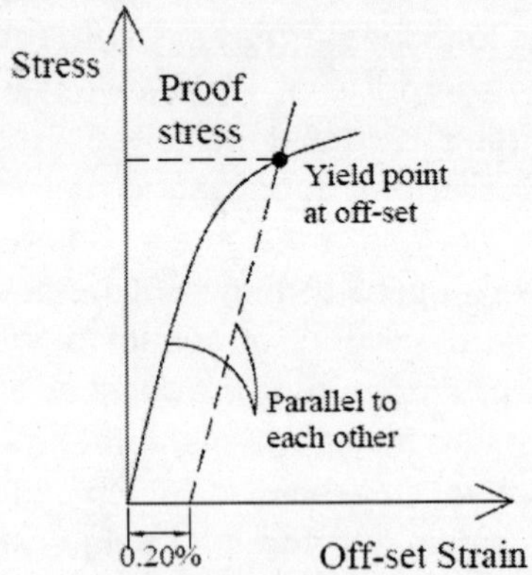

Fig. 3.11: Tensile test curve of brittle materials

A typical cross-section of a failed brittle material is shown in Fig. 3.12. Brittle materials fail by two mechanisms as discussed below:

Trans-granular fracture: The fracture plane travels through the grains of the materials. Hence, the plane also changes directions, due to different lattice structure and orientations of the grain atoms. These cracks choose the path of least resistance, which is visible physically as a bumpy cracked surface.

Inter-granular fracture: In this type of failure, the failure plane travels along the grain boundaries, not through the grain body. If the structure is assumed as a puzzle, this crack travel along pre-cut edges (already existing weaker zones).

Fig. 3.12: Cross section of brittle materials during failure (https://sm-nitk.vlabs.ac.in/exp/tensile-test-cast-iron/theory.html)

One of the characteristics of a brittle failure is that the two broken parts can be reassembled to produce the same shape as the original component as there will not be a neck formation like in the case of ductile materials.

Ductile materials:

A material is told to be exhibiting ductile properties if the post elastic strain during tensile testing is more than 5%. Post elastic strain is different from the plastic strain in a way that the plastic strain is observed up to the region of plastic deformation, which the post-elastic strain is seen up to the fracture point, which include the plastic strain as well.

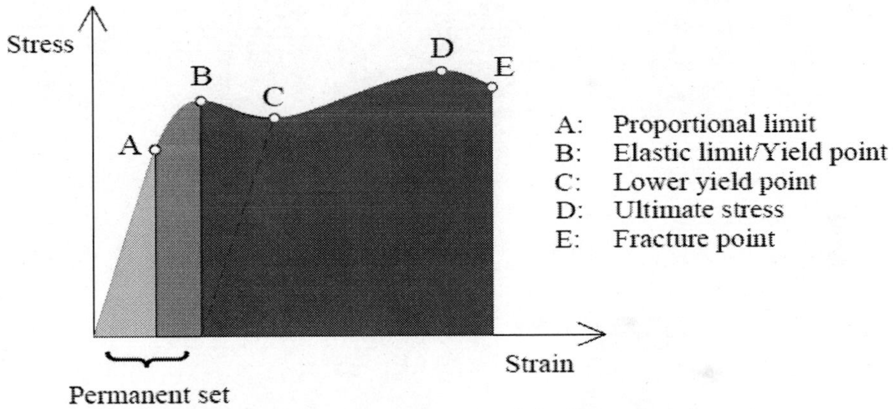

Fig. 3.13: Tensile test curve of ductile materials

The above Fig. 3.13 shows a typical stress-strain curve obtained for a mild steel section, which is ductile in nature. The entire curve can be divided into four zones with following stress points as follows:

Elastic limit: The material obeys Hooke's law up in this zone and the material when unloaded, regains its original shape. In this region, the material behaves elastically in nature.

Proportionality limit: It is the maximum limit up to which the Hooke's law is valid, hence the linearly elastic property of the material is also valid up to this point Hence, this region, up to proportionality point is also called linearly elastic region.

Yield point: It is this point where the elastic property diminishes and plastic region starts to appear. It indicates the start of the zone where minor or no increment in loading causes large deformations in the material, the reason being that number of dislocations (voids and cavities in the crystal structure) in the crystalline lattice increases. However, also the dislocations are increasing, the stress induced on the material is also increasing. Hence, the material also hardens, but up to a limit, called the ultimate limit.

Ultimate limit: Till this limit, the material can sustain maximum amount of stress induced in it.

Fracture limit: Once, the stress induced becomes more than the ultimate limit, a cup and cone shaped formation appears, due to large number of dislocations, which reduced the effective area of cross-section to withstand the force induced. Finally, the material gets torn apart at the weakest section, in the form of cup and cone shape. This limit is called the fracture limit.

A failure patter of a typical mild steel specimen is shown by a schematic diagram depicted below in the Fig. 3.14.

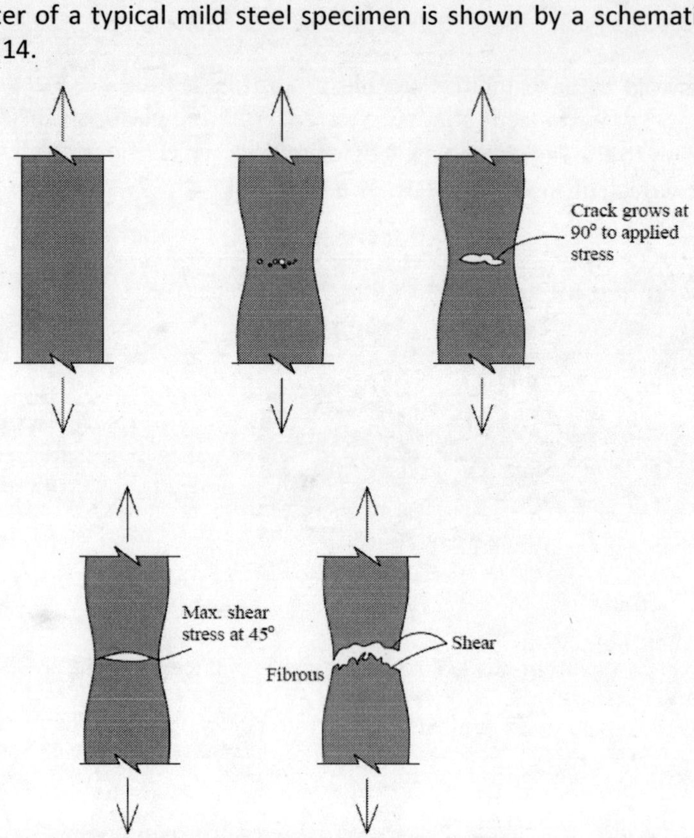

Fig. 3.14: Cross section of ductile materials during failure

3.5.2 Compressive strength

Compressive strength is determined using the compression test. Similar to tensile test the test can performed using universal testing machine or a similar apparatus. Hydrostatic piston moves to produce a compressive strain on the material. And the load cell and strain gauges are used to measure the stress vs strain curve of material. In compression test the phenomenon of plastic deformation is not visible, and hence ductile and brittle material create a similar stress-strain curve. However, it should be noted that yield point is visible in compression test of ductile materials. Compressive strength is largely identified as the ultimate failure strength or maximum strength of the material. Yield strength in case of brittle materials is defined in terms of predefined strain value. In compression test, slender members ($L \gg D$) can also undergo buckling failure. Since in most standard testing protocols slender members are not used during the test buckling is not observed.

3.5.3 Flexural strength

Flexural or bending tests are used to determine the elastic modulus, shear modulus check deflection and determine flexural strengths. They differ from pure tensile or compressive tests in a way that the stresses and strains induced due to such loading are not uniform over the cross section. Hence, stress-strain curves cannot be predicted directly based on experimental data.

When a slender member is subjected to a loading transverse to its longitudinal direction, the member bends in the direction of the applied load. This load applied produces the flexural stress and resistance to such bending is called flexural strength. This test is used to determine the tensile/rupture properties which are caused in a member due to loading at any transverse direction, specifically for brittle materials. It is because brittle materials have low tensile strengths and when a tensile force will be applied, they will crack from the grips of the UTM. In general, this test is done by four-point loading method. Sometimes, three-point loading system is also used to determine flexural strength for specimens of rectangular cross-sections, with smaller spans. Bending stress at any depth of cross-section is determined using the pure bending equation, also called circular bending equation:

$$\frac{M}{I} = \frac{\sigma}{y} = \frac{E}{R} \qquad (3.12)$$

When four-point loading system is involved, where the beam is supported on ends by two supports and two-point loads are acting at equal spacing on the span, a zone of pure bending is generated as shown in the following Fig. 3.15. In this method, any shape of cross section of beam may be used.

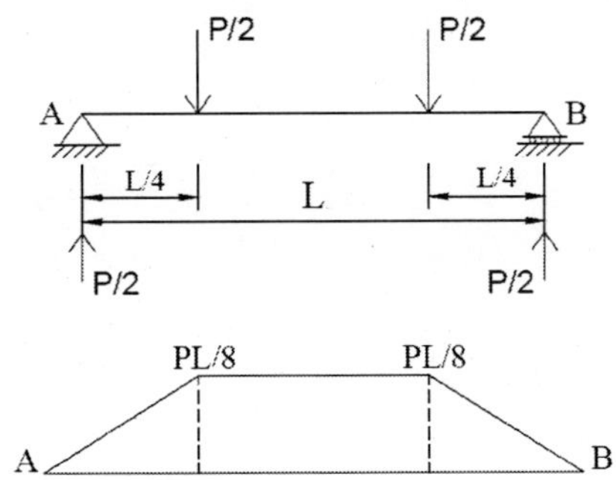

Fig. 3.15: Pure bending condition by 4-point loading system

3.5.4 Torsion strength

When a twisting force is applied at any cross-section of the member, a moment, called the torsional moment is generated, which induces a stress called torsional stress. Practically, torsional stress may be generated from transverse loading as well. However, it is different from bending in a way that when the transverse loading is acting anywhere except on the shear centre of the section, the twisting will occur, else no twisting generates.

Torsion test is conducted on a circular bar placed in the torsion testing machine (Fig. 3.16) in such a way that its longitudinal axis coincides with the axes of the grips holding the specimen. Other cross sections are not used due to the problem of change in cross-section while twisting them. Then, one of the grips is rotated at a constant speed until the test piece breaks, rotation being measured by a troptometer. The torsional strength is determined using pure torsional equation as given below.

$$\frac{T}{J} = \frac{G\theta}{L} = \frac{\tau}{r} \tag{3.13}$$

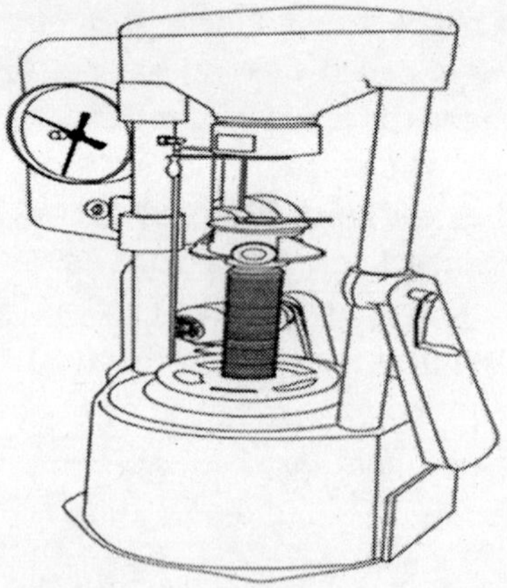

Fig. 3.16: Torsional testing machine
(https://sm-nitk.vlabs.ac.in/exp/torsion-test-mild-steel/theory.html)

The various failure patterns due to torsional stresses are depicted in the following Fig. 3.17.

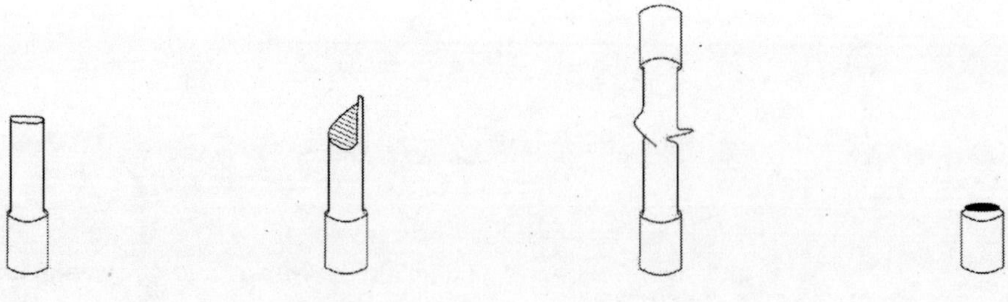

(a) (b) (c) (d)
Fig. 3.17: Different types of failure patterns during torsional test

- **Ductile failure:** failure occurs in ductile materials on plane of maximum shear stress, perpendicular to the specimen axis.

- **Brittle failure:** failure propagates on a plane of maximum tensile stress, 45° to specimen axis, creating a helical failure plane.
- **Buckling failure:** observed in hollow ductile pipe
- **90° failure:** observed in plane in hollow brittle pipe

3.5.5 STRENGTH OF CERAMIC

The resilience of ceramic materials at quite high temperatures is one of their standout qualities. Many of them are stronger than metals at significantly greater temperatures. Although most ceramics have exceptional high temperature strength, some are vulnerable to thermal shock. Ceramics have characteristics that are very different from either those of metals or organic polymers. Although they have a strong compressive strength, they are quite weak in tension. They are good thermal and electrical insulators, resistant to oxidation and corrosion, resistant to high temperature creep, strong, hard, and brittle, and possess a high melting temperature.

Mechanical properties of ceramic phases

- **Compressive strength**: the compressive strength of ceramics is much higher than the tensile strength. The fact that materials like cement, glass, and brick are always employed in compression rather than tension makes this obvious. Due to their great compressive strength, tempered glass is utilised for glass doors, and other high strength applications.
- **Tensile strength:** Although ceramics' tensile strength is high in theory, it is typically rather low in practise. Stress concentrations at the pores, grain corner, or microcrack are often the cause of failures. For instance, pores rapidly reduce a material's strength because they allow for the build-up of tension. Since loads are distributed across the crack and not concentrated at the root, this stress concentration has no impact on compressive strength. On the other hand, glass fibres have extremely high tensile strengths, frequently exceeding 700 MN/m^2.
- **Shear strength:** With the exception of a few materials, like clay, ceramics typically have high shear strengths and low fracture strengths. As a result, they frequently fail as a brittle failure, which is referred to as failing nonductilely. Below the softening temperature, non-crystalline ceramics are always brittle. As a result, brittle cleavage along specific crystallographic planes causes the failure.
- **Plastic deformation**: Most materials have restricted slip, which prevents plastic deformation. However, the assumption that ceramic materials have stronger slip resistance than metals is specifically excluded from clays and other layered structures. Because slip is far more likely to happen in ceramic materials after such deformation, these materials go through plastic deformation. On the other hand, rather than slip, viscous flow causes plastic deformation of glass.
- **Torsional strength.:** Torsional strength is rarely regarded as a crucial characteristic of ceramics because tensile and cantilever requirements will reveal the material's torsional strength.
- Impact factors in some circumstances, the majority of ceramic materials must tolerate dynamic forces. The impact values for vitrified ware, dense stoneware, and steatite vary from 0-72 to 1-22 Nm, 1-12 to 1-33 Nm, and 16 to 2-8 Nm, respectively.

- **Modulus of elasticity:** The strength of the bond is shown by the ceramic materials' high modulus of elasticity, which ranges from 7×10^{10} to 42×10^{10} N/m^2.
- **Transverse strength:** As it is hard to determine transverse strength in ceramic materials, ceramics are not employed in applications where transverse strength is a crucial factor.

Electrical properties of ceramic phase:

The composition, texture, size, and density of a material, as well as its temperature and time, all affect its electrical properties. These elements have a significant impact on how a ceramic material behaves electrically. Ceramic materials are employed as dielectrics, insulators, conductors, and semiconductors. Additionally, ferroelectric and piezoelectric materials made of ceramics are employed.

Because the electrons connected to the atoms that make up ceramics are shared through covalent or ionic interactions, ceramics are often poor conductors of electricity. The following variables are the key influences on ceramics' electrical properties:

- Dielectric strength
- Volume resistivity
- Dielectric constant
- Dissipation factor (or loss factor)

A known volume of insulating material's volume resistivity serves as a gauge of that material's resistance. The volume resistivity of ceramics diminishes as temperature rises. The voltage required to break down an insulator and for it to conduct electricity is measured by the insulator's dielectric strength. The ceramic loses dielectric strength due to moisture absorption, impurities, surface polish, and many other reasons, making it a poor insulator. The ability of an insulating material to store electrical energy is determined by the dielectric constant. This attribute needs to be high if the ceramic is going to be utilised as a capacitor. High dielectric constant is a property of barium titanate. It is preferable for a ceramic to have a low dissipation factor when it is utilised as an insulator in an alternating current field. The influence of temperature, impurity concentration, porosity, and frequency on this dissipation factor is considerable.

- **Ceramic insulator**: Both the electrical insulators and the functional components of an electrical circuit are made of ceramic materials. Ceramics only need to be electrically inert and able to isolate two conductors with differing potentials in order to be utilised as insulators. The electrical field and the charges inside the ceramic materials' structure must interact when they are used as functional parts. Materials that are thought of as insulators can rupture under high voltage
- **Dielectric ceramics:** Ceramic materials are employed as condenser dielectric medium and as transformer cores because of their high dielectric capacity. High-grade fire clays and porcelain have strong dielectric properties. Temperature affects the dielectric. The most popular ceramic materials have dielectric constants of up to 12, including glass, porcelain, alumina,

quartz mica, and asbestos. Some ceramics, like the mineral rutile, have greater dielectric constants (TiO_2).

- **Ceramic conductors**: Movement of free ions and electrons determines a substance's conductivity. Free electrons or ions with an electronic imbalance cannot move in ceramic materials. Therefore, at lower temperatures and particularly in freezing situations, they are effectively non-conductors. Ceramic materials have a high resistance, which causes them to heat up quickly. Ionic diffusion and mobility increase with increasing temperature because heating them causes the ionic bonds to become disturbed. With more atomic structure flaws, mobility also rises. At low temperatures, Ion movement becomes incredibly tiny, but conductivity is still detectable because of the impurity ions that carry the charges. Normal glass and silicates are insulators, but when they are molten, they become excellent conductors.
- **Ceramic semi-conductors:** Ceramic compounds are typically insulators, but if they contain multivalent transition elements, they can change into semi-conductors. Semi-conductors are substances that exhibit conductor-like characteristics at high temperatures and insulator-like characteristics at low temperatures. Germanium (Ge) and silicon (Si) are both frequently used semi-conductor materials

Thermal properties of ceramic phases

- **Thermal conductivity:** Ceramic materials lack enough electrons to achieve electronic thermal conductivity. Here, the heat is transferred by the interplay of the structural movement and thermal conductivity, whereas at high temperatures, heat is transferred through the transfer of radiant energy. Temperature, porosity, and impurity content all contribute to a reduction in thermal conductivity. Maximum density, which is absent in the majority of ceramic materials, is necessary for maximum thermal conductivity.
- **Thermal capacity**: clay bricks have a specific heat of 0-25 at 1000°C and 0-297 at 1400°C, respectively. The specific heat of carbon bricks is around 0-812 at 200°C and 0-412 at 1000°C.
- **Thermal shock:** Many ceramic compositions contain lithium compounds, which minimise thermal expansion and offer superior thermal shock resistance. Common ceramic materials are hot pressed silicon nitride, fused silica, cordierite, zircon, silicon carbide, beryllia alumina, porcelain, and steatite, which are ranked in decreasing order of thermal shock resistance.
- **Porous Ceramics as Thermal Insulation**. The amount of loosely bound electrons in ceramic materials is insufficient to produce electronic heat conductivity. Here, the heat is transferred by the interplay of the lattice's motion and particle conductivity, but at high temperatures, the heat is transferred through the transfer of radiant energy. energy Ceramics' porous nature reduces heat conduction by reducing interaction between atoms and molecules. Additionally, air is trapped in between pores if porosity is higher. Because it conducts heat poorly, the air hinders convectional heat transfer. Because of the aforementioned factors, thermal insulator ceramics are porous.

3.6 CREEP

3.6.1 Fundamentals of creep

It's a frequent misconception that elastic and plastic strains emerge instantaneously when tension is applied. A creep is further deformation that happens gradually over time. In engineering design, creep is frequently crucial, especially in applications involving high temperatures, like steam, nuclear reactors, jet and rocket engines, and turbines in power plants. Other instances of creep include the gradual loosening of plastic eyeglass frames, failure of lightbulb filaments, and slow deformation causing the plastic pipe to rupture as well as glacial ice movement. The amount of creep deformation in crystalline ceramics and metals is significant only at temperatures that are typically between 30% and 60% of their absolute specific material freezing point. Even at ambient temperature, creep is frequently a problem for polymers that are in a leathery or rubbery state. Concrete creeps at ambient temperature, but the mechanism slows down over time so that after the first year or so, only minor new strains appear.

A creep-sensitive design is likely to be impacted significantly by the choice of material. Engineering metals that are subjected to high temperatures typically contain alloying elements like chromium, nickel, and cobalt, with the proportion of these expensive components rising with temperature resistance. Even better temperature resistance is possible with new, robust ceramic materials than it is with the greatest metal alloys. The temperature resistance of polymers, on the other hand, is highly constrained. As chemical activity rises with temperature, other environmental impacts including oxidation and environmental cracking are also likely to be problematic. The combination of cyclic loading and time-dependent deformation is another difficulty that frequently appears in engineering settings. There could therefore be a negative creep-fatigue relationship that quickens the fatigue process. Tools for a design that can be utilized to prevent failure due to creep are provided by the engineering approaches that have been developed for studying and predicting creep behaviour. Excessive deformation is a worry. The material can separate (crack) as a result of the creep process, which is another possibility.

Creep testing

Applying a steady axial force, either in tension or compression, to a bar or cylinder made of the test material is the most used method of creep testing. Simple dead weights and a lever system, as depicted in Fig. 3.18, may be utilised to hold the force constant for extended periods of time. When a rupture occurs during the test, the time at rupture is recorded in addition to the creep strain over time. Tests on a particular material are often conducted under a variety of loads and temperatures, with test times ranging from a few seconds to several years.

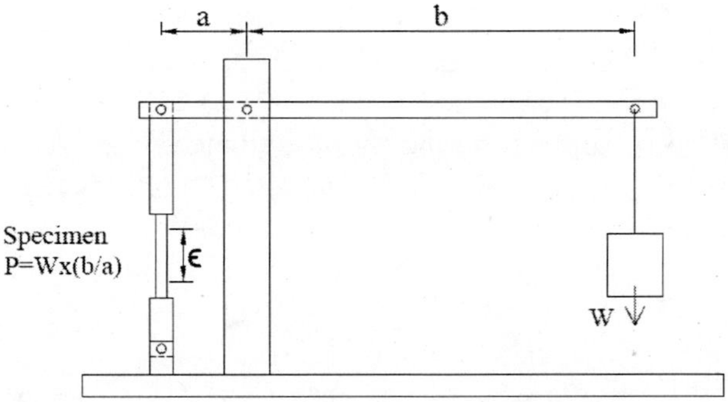

Fig. 3.18: Schematic diagram of creep testing machine

3.6.2 Characteristics

The behaviour observed on a graph of strain versus time is usually similar to Fig. 3.19

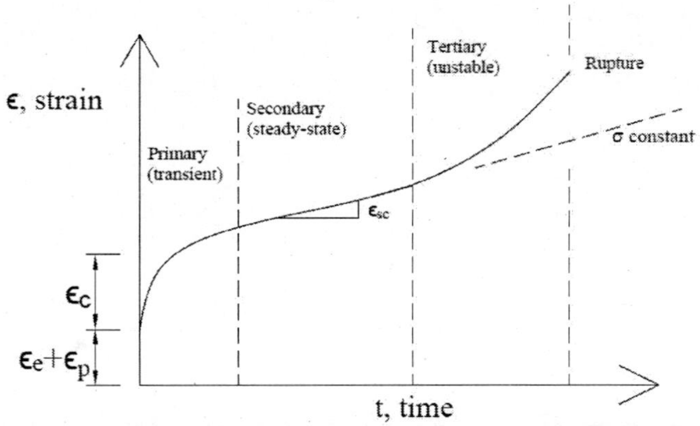

Fig. 3.19: Strain versus time diagram during creep under constant force

For several kinds of materials, there are distinctly different physical mechanisms that cause creep. Additionally, distinct mechanisms operate at varied stress and temperature combinations even for a given material. A solid material's internal movements of its atoms, vacancies, dislocations, or molecules all depend on time and happen more quickly at higher temperatures. Such motions fit into the general category of behaviour known as diffusion and are crucial in describing creep behaviour.

By developing cracks, fatigue, or other damage as a result of the extreme strain, creep deformation can continue until the material ruptures. For instance, voids may develop in crystalline materials through a process known as creep cavitation along grain boundaries or at other sites of localized stress concentration, such as precipitate particles. The expansion and merging of grain boundaries or other voids then result in fractures, which may develop into creep rupture, a type of fracture. However, the process of dynamic recrystallization, in which these voids are basically mended as they try to form, can take place if the temperature is high enough in a ductile and reasonably pure metal. The potential for large deformations increases and failure eventually happens by necking. Large uniform or necking deformations are frequently seen prior to creep rupture in ductile polymers. When creep is present in an engineering design, there must be no excessive deformation or rupture during the intended service life, which is usually rather long—perhaps 20 years or more. Creep data, however, is often only available out to 1000 h (42 days), or occasionally 10,000 h (14 months), but seldom to 100,000 h due to test-time limits (11 years). One such method is to extrapolate the suitable stress versus strain curve to estimate creep strains for the service temperature of interest in order to predict the behaviour at low strain rates and long periods.

Creep is frequently involved in practical applications at high temperatures, and these phenomena may work in concert. Because of the complex combination of thermal stresses brought on by significant temperature variations combined with cyclic loading as the aircraft flies at a steady speed, changes speed, lands, and halts the engines, etc., various components of aircraft jet engines, for instance, experience periods of both varying and steady stress. The combined creep and wear of high-temperature components in nuclear reactors and different pressure containers is another issue. Cycling frequency is crucial where creep and fatigue combine because slow frequencies give creep more time to cause damage. The frequency-modified fatigue technique, devised by L. F. Coffin, is one strategy created with such effects in mind. The correlations between cyclic stress, strain, and strain life are generalized, making the different material constants functions of temperature and frequency. Stress-life curves and time-temperature parameters can still be used to determine life if stresses vary seldom. However, a more complicated situation develops that necessitates specialized study if stress changes happen so frequently that the cyclic loading starts to induce fatigue damage.

3.7 IMPACT

In strength test the load increments are gradual and in creep the load application is very slow or constant. It is possible that a material may undergo sudden application loads. This sudden application is known as impact loading. Impact loads acts in cases, like crash barriers, where the load transfers energy to the material. The ability of material to resist impact loading is defined as impact strength. The amount of energy required to break a standard size of material is used to define its impact strength. The energy at which the material fails is known as fracture energy and the corresponding strength is known as fracture toughness.

In a typical impact application, the applied load causes both stress and strains. The area under the stress strain curve represents the energy per unit volume of the material. If the sudden developed strains exceed the strain at failure, the material fails. In case of ductile materials, like steel, plastic deformations allow for higher strain values. And thus, ductile materials have significantly higher impact strength.

The impact energy of any material can be measured by any one of the following approaches:

- A high value of impact load is applied on a sample of standard size to cause its failure. The residual energy of the load is measured after the failure of the sample. The amount of energy absorbed by the material during failure is defined as its fracture energy. Charpy and Izod tests are example of this type.
- A low value of impact load is applied on a sample, repeatedly, till the sample fails. The value of load should be high enough to cause plastic strains. After multiple impact loadings the plastic strains accumulate and the material fails. The total energy required for the failure of material, defined in terms of number of times impact load is applied, shows the impact strength of material. Ball drop test is an example of this type.
- An impact load is applied only once on a sample and its stress and strains are measured. The area under the stress-strain curve indicates the amount of energy absorbed by the material. This type of impact test can measure impact energy at different failure conditions. Drop impact test is an example of this type.
- Stress-strain curve is plotted in a typical strength test. The area under the strength test, till the point of ultimate failure is used to denote fracture toughness of the material.

Brittle fracture of steel temperature transition approach

Steel is one of the most commonly used metal in engineering materials. Application of steel ranges from large scale applications, like steel bridges, to small scale applications, like steel frames for windows. Steel for most conditions behave like a ductile material, however at low temperatures steel starts to behave like a brittle material. This brittle nature of material, or the absence of plastic deformations, results in lower impact strength. The temperature below which steel starts to behave like a brittle material as opposed to ductile material is known as *ductile-brittle transition temperature* or *nil ductility transition temperature*.

The ductile-brittle transition temperature is defined as a temperature below which fracture energy falls below a predetermined value. This ductile brittle transition is observed in alloys like low carbon steel. It is not necessary for all engineering materials to exhibit a distinct transition between ductile and brittle behaviour, like FCC lattice copper alloys. On the other hand, some materials, like with BCC lattice, undergo a very sharp temperature transition.

During civil engineering applications in cold climatic regions, it is possible that the external temperature drops below the transition temperature. This also possible in certain industries where coolants and similar chemicals are being kept. At temperatures below the transition, steel with an acute crack can fail brittlely if the crack is wide enough and the applied stress is strong enough. This type of failure has been observed in case of Titanic. In case of steel, presence of alloying metals like nickel and manganese can lower the transition temperature, below the operational temperature. Thus, by lower the transition temperature one can ensure that actual temperature does not fall below the same and prevent brittle failure.

This phenomenon of ductile brittle transition can also occur due to nuclear radiations, typically in nuclear power plants, and is known as embrittlement. Embrittlement results in an increase in the transition temperature, and may increase the transition temperature above operational conditions. Therefore, special attention must be given to the transition temperature when applying engineering materials, like steel, in nuclear power plants.

3.8 CONCEPT OF FATIGUE OF MATERIALS

Material fatigue is a phenomenon that occurs when structures are subjected to a cyclic load. Even when the experienced stress level is well below the static material strength but with repeated cycling, the microscopic damage may build up until it becomes a crack or other macroscopic damage that causes the component to fail. Once a fracture has started, each loading cycle will cause it to propagate a little amount, even though the repeated alternating or cyclic loads are significantly lower in intensity than the usual strength. Fatigue is the most common cause of mechanical structure failures. Metals, polymers, and ceramics, which are all frequently used industrial materials, are all prone to unexpected fatigue failures.

Despite being brittle, the fracture may take a while to develop depending on the strength and frequency of the cyclic stress. However, if the break is not spotted, there is little or no warning before failure. At given peak stress, the number of cycles needed to cause fatigue failure is often relatively high, but when the stress is increased, the number of cycles required reduces. As long as the peak stress, which is also referred to as the fatigue strength, is below the endurance limit value, cyclical stresses can be applied to some mild steel indefinitely. Breaking a thin wire with the hands after repeatedly bending it back and forth in the same spot is a good illustration of fatigue failure.

The three steps that characterize the fatigue failure process are as follows:

- Fracture initiation; is the process by which a hairline crack develops at a site of high-stress concentration.
- Crack propagation, in which each stress cycle causes the crack to propagate a little bit. The phase of crack growth often consumes the majority of the fatigue life.
- Ultimate failure; once the developing crack reaches a crucial size, ultimate failure occurs quickly.

Stress vs Life (S-N) Curves

A fatigue fracture or other type of damage will form on a test specimen of a material or engineering component when it is subjected to a considerable high level of cyclic stress, ultimately causing the member to fail. The number of cycles to failure may be lower if the test is repeated under more stress. A stress-life curve, also known as an S-N curve, can be created by plotting the results of such testing at various stress levels. The majority of fatigue tests are performed at a "Constant Amplitude Stress Level," meaning that the maximum and minimum stresses are constant throughout the test. The applied stress, S, versus several cycles to failure, N, or say S-N curve for a material, is the way of representing the results of fatigue testing. On semi-log or log-log graph paper, S-N curves are typically plotted, with each plotted point representing the outcomes of a single test specimen. To provide enough data for the interpretation of the tested material's fatigue behaviour, the fatigue test is often conducted using a minimum of 8 to 12 specimens.

Factors Influencing Fatigue

The factors like stress concentration, specimen size, cumulative fatigue, surface condition; and change in stress level, metallurgical variables, corrosion, and temperature, in addition to the characteristics of the applied stresses like maximum stress, mean stress, and stress ratio, significantly affect the fatigue behaviour of the materials. Since the starting of a fatigue crack typically occurs close to the surface, rough surfaces should be avoided since they might cause stress concentration, which can cause further propagation of fatigue crack and ultimately catastrophic failure takes place. High service temperatures and corrosive environments are considered to have a negative impact on a material's fatigue properties because they speed fatigue initiation and propagation rate.

3.9 FRACTURE MECHANICS

Fracture mechanics is the scientific study of the growth of cracks in solid materials. The force needed to cause cracks is calculated in this study using analytical solid mechanics methods, and experimental work is done to determine how much resistance the material offers to fracture.

The rapid fracture may happen with static loading, but it won't change with time. Brittle fractures are those that happen with little to no plastic deformation. The glass and other materials that are similar to it can fracture in this way. When there is a sudden application of force, such as during impact loading, brittle fracture is more prone to happen. When cracks or other severe defects are present, the brittle fracture can happen even in materials that deform significantly plastically (such as ductile steels). Temperature, strain rate, and stress level are the three variables that have a significant impact on the kind of fracture. Steel undergoes brittle fracture at low temperatures, which can happen rapidly and without prior notice. Therefore, a brittle fracture can start with the proper combination of low temperature, a sudden change in section size or a defect/crack, and also the existence of tensile stress. The ability of any material to withstand brittle fracture is indicated by the material property known as fracture toughness.

A.A. Griffith, a British aeronautical engineer, was the first to create a theory of fracture mechanics that applied notions from elastic strain energy to elastic materials. Griffith proposed that glass's low fracture strength is caused by microscopic defects, and in order to test this theory, he intentionally put a defect into his experimental glass specimens (as seen in Fig. 3.20(a)). He discovered that the square root of the defect length (a) and the fracture stress (f_f) produced a virtually constant product.

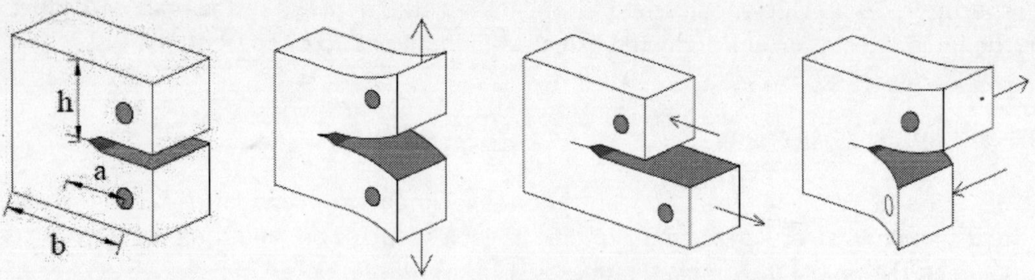

(a) Sample (b) The three fracture modes

Fig. 3.20: Fracture-toughness test and fracture modes

$$f_f = \sqrt{\frac{2E\gamma_s}{\pi a}} \quad \text{and} \quad K_c = f_f\sqrt{\pi a} \quad (a \ll b) \tag{3.14}$$

Where f_f is the fracture stress, E is the Young's modulus, a is the crack length in m, γ_s is the surface energy density of the material (J/m2). The Griffith equation only applies to perfect brittle materials like glass and is highly reliant on the crack size a. Additionally, because it is difficult to calculate surface energy γ_s, Griffith's criterion is rarely applied practically.

Although the energy balance approach offers significant insight into the fracture process, it has been shown that a different approach that directly investigates the stress condition close to the tip of a sharp crack is more practical in engineering practise. Fig. 3.20(b) depicts three different types of cracks, denoted as modes I, II, and III. Mode I is the important normal-opening mode, whereas Modes II and III are the shear sliding and tearing modes, respectively, which are not frequently taken into account.

When the material is more ductile, like steel, taking into account the surface energy alone is unable to produce a realistic fracture model. Following equation has been proposed for such materials where there are plastic deformations.

$$f_f = \sqrt{\frac{EG_c}{\pi a}} \quad \text{with} \quad G_c = 2(\gamma_s + \gamma_p) \tag{3.15}$$

With fracture toughness $K_c = \gamma \times f_f \times \sqrt{\pi a}$

Where, γ_s is the surface energy, γ_p is the plastic dissipation per unit area of crack growth. γ=1 for a centre double-ended cracked plate (a/b $\leq$ 0.4), γ=1.12 for a single-ended-cracked plate (a/b $\leq$ 0.13), and $\gamma=2/\pi$ for an embedded penny-shaped circular crack of radius a in an infinite plate.

It is common for materials with high strengths to also have low fracture toughness, and vice versa. K_c can be evaluated for various materials using standard testing procedures. In actuality, K_c is also influenced by the specimen geometry and the loading circumstances, and it is particularly sensitive to the specimen thickness. K_c is a feature that depends on the microstructure. Only under conditions of plane strain can data for fracture toughness be consistently and reliably reproduced.

The quantities G and K_c have the following relationships.

$$G = \frac{K_c^2}{E} \tag{3.16}$$

Where $E' = E$ for plane stress, $E' = E/(1 - \nu^2)$ for plane strain, and ν = Poisson's ratio

The plastic deformation that typically goes along with ductile fracture may include progressive tearing. During designing pressure vessels and massive welded structures like bridges and ships, fracture mechanics and brittle versus ductile fracture must be taken into account. In contrast to ductile fracture, creep rupture is a time-dependent phenomenon.

Sometimes a combination of chemical and mechanical forces leads to fracture. Stress corrosion cracking is the term for this type of fracture, which is typically seen in chemical industrial structures. Additionally, in the presence of chemicals like sodium hydroxide, some low-strength steels are susceptible to cracking. When tensile stresses are applied to prestressed concrete, atomic hydrogen may be produced as a result of the action of acids, which can penetrate the steel surface and make it brittle and prone to fracture. High-tensile steel wires used in prestressed concrete can suffer significant tensile strength loss from even a modest amount of hydrogen. Hydrogen embrittlement can also result through the usage of cement high in alumina and cement made from blast furnace slag. Hydrogen embrittlement also occurs when high-tensile steel wires are kept with incompatible metals like zinc and aluminium. When in touch with high-tensile steel wires in the presence of moisture, even minute amounts of sulphur can significantly lessen their strength. Steel needs to be carefully protected against acid action to avoid hydrogen embrittlement. To lessen the likelihood of contamination, protective covering like bitumen crepe-paper should be applied during shipment. Always store the wires in a dry environment.

3.9.1 Fracture toughness testing

The fracture toughness of materials can be assessed using a variety of tests, most of which make use of notched specimens with different configurations. To assess the fracture toughness of metallic materials across a variety of temperatures to calculate design life and crack growth, or residual life, the linear-elastic plane strain fracture-toughness test is utilised. Other materials, like polymers and ceramics, which are covered by ASTM D5045 and ASTM C1421, respectively, also undergo comparable tests. In these materials, as specimen thickness is increased, the K_K value decreases.

A tiny crack is pre-set into the test specimen prior to applying the load in the three-point beam bending test, which is another frequently used test (see Fig. 3.21). Plotting the force versus displacement (P-δ) behaviour reveals the crack's growth. The PQ relating to the preliminary phase of cracking is identified by a departure from the linearity of the P-δ plot. For this location, K_K is calculated.

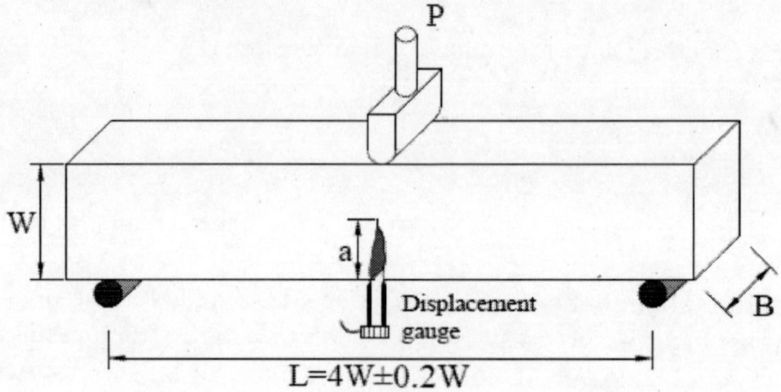

Fig. 3.21: Fracture toughness test on bend specimen

The physical opening of the fatigue crack tip at the point of failure is measured using the crack tip opening displacement test, or CTOD test. It is frequently used to gauge the toughness of materials with lower strengths since they only show a little plastic deformation before failing. All CTOD testing are conducted in the -129°C to 200°C temperature range. The Charpy impact test is a frequently used standard test procedure that involves subjecting a standard rectangular simply supported beam with a V-notch or a U-notch at mid-length to an impact load from a swinging pendulum and measuring the energy absorbed by the specimen.

3.9.2 Structural integrity of assessment procedure and fracture mechanics

Structural Integrity Assessment is a process to determine if a structure is prepared to resist operating circumstances safely and reliably for the duration of its anticipated lifetime. For the duration of its anticipated lifetime, this involves bearing its own weight and attempting to avoid deformation, breaking, and catastrophic failure.

Causes of Structural Failure

Manufacturing Error: This involves utilizing the incorrect materials or failing to adhere to manufacturing norms or processes. It may also be the result of subpar construction, out-of-tolerance components, etc.

Weak Structure: The building is not sturdy enough to support the load that is placed on it. This is typically caused by improper geometrical design or material selection.

Defective Material: Using incorrect materials or deviating from manufacturing norms or procedures are examples of this. Additionally, it may be brought on by subpar construction, out-of-tolerance components, etc.

Adverse Environmental Considerations: When it comes to specific environmental situations, such natural disasters, engineers may disregard mitigating factors.

SINTAP Procedure (SINTAP – Structural Integrity Assessment Procedure)

The SINTAP procedure is based on principles of fracture mechanics. The following tasks of a fracture mechanics analysis are made possible by this principle:

- Crack detection in a component during service.
- A component can be set-off for a potential crack during the design phase. The crack's dimensions and location must be designed so that it can be identified by non-destructive testing (NDT) during the final quality control or while the component is in use.
- Determination of critical crack dimensions for subsequent NDT testing.

Fracture Mechanics

Failure theory in fracture mechanics

- uses energy criteria to assess material failure, sometimes in conjunction with strength (or yield) criteria
- believes that breakdown is spreading throughout the structure rather than happening all at once in the failure zone or surface.

Failure in fracture mechanics:

$$stress + crack\ geometry > material\ fracture\ toughness$$

Failure in strength of materials:

$$stress > material\ strength$$

Design, material selection and alloy development, and assessing the importance of faults are three key applications of fracture mechanics. Monitoring and control, as well as failure analysis, are ancillary areas.

Consideration of fracture mechanics is crucial for a number of reasons:

- A part may already have cracks or they may form as a result of extreme stress or exhaustion.
- The failure of parts at loads below those predicted by a strength-of-materials method can result from disregarding fracture mechanics.
- Usually, a material's fracture toughness reduces as its strength increases. Many engineers can end up in danger because of their tendency to favour stronger materials.
- Brittle fracture failures are sudden, devastating, and come with little forewarning.

3.10 OTHER MECHANICAL CHARACTERISTICS

The previously discussed mechanical characteristics, commonly describe the failure of materials. There are other mechanical characteristics as well, which, can be used to describe the behaviour of engineering materials. Examples for the same can be hardness and internal friction. These tests help in understanding the performance of material in specific applications. For example, a structure in flowing river may experience surface damage in the form of abrasion. This surface damage can cause change in shape and directly contribute to load-based failures. Therefore, it is important to appreciate theses non-load-based mechanical characteristics from application point of view.

3.10.1 Hardness test

Hardness of any material is the resistance of the material against abrasion, wear and tear, scratches and cuttings against external forces. It is determined using machines like the Rockwell and Brinell testing machines, as shown in Fig. 3.22. In these machines, an indentation is made on the specimen and the resistance against the indentation is measured by determining the force induced by the machine and the cross section of the worn-out portion of the specimen.

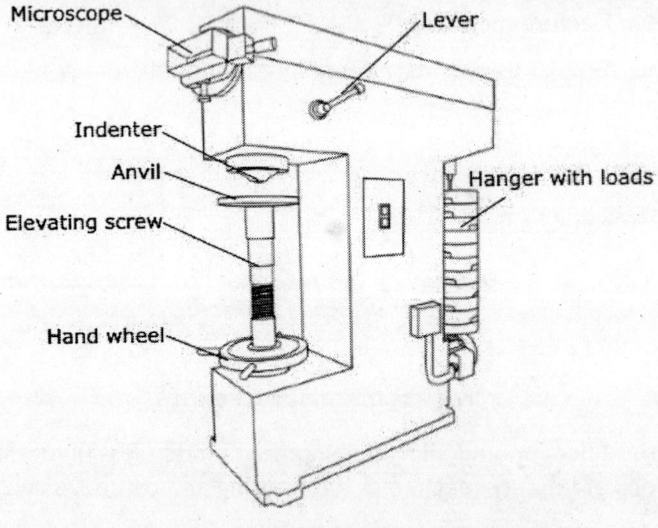

Fig. 3.22: Hardness testing machine
(https://sm-nitk.vlabs.ac.in/exp/brinell-hardness-test/theory.html)

Brinell testing machine is used when the material to be tested has a rough or coarse structure, which cannot be tested using other methods. For example, castings and forgings. However, the surface of the specimen needs to be smooth, clean from any foreign matter and lubricants. In this method, an indenter made of a hard-metal ball of diameter D (in mm), as shown in Fig. 3.23, is forced into the specimen surface by a force P (in N) and the diameter of the indentation d (in mm) is measured and converted into curved surface area of the intended surface. The test piece standards and testing procedures must comply with the standards laid down on IS 1500 (2005).

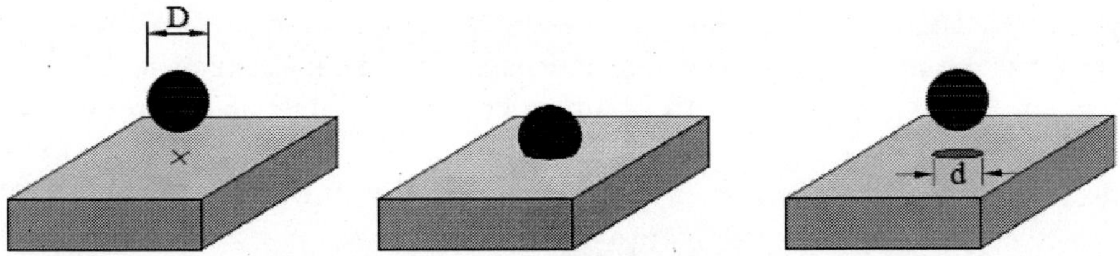

Fig. 3.23: Brinell indenter and depth of indentation

The following equation is used to determine the hardness of any specimen using Brinell testing method.

$$\text{Brinell hardness} = \frac{2P}{\pi D(D - \sqrt{D^2 - d^2})} \tag{3.17}$$

In case of Rockwell testing method, the same Brinell testing machine may be used. However, the intender here is made of hardened steel or diamond cone, apex angle of which is 120° and radius of curvature at the tip being 0.2mm. Hence, the diamond cone indenter is used for harder metals like carburized steel, cemented carbides etc, as shown in Fig. 3.24. Sometimes, ball indenter may also be used, in case of soft metals like copper, aluminium, brass and gray cast iron. The test piece standards and testing procedures must comply with the standards laid down on IS 1586 Part 1 (2012).

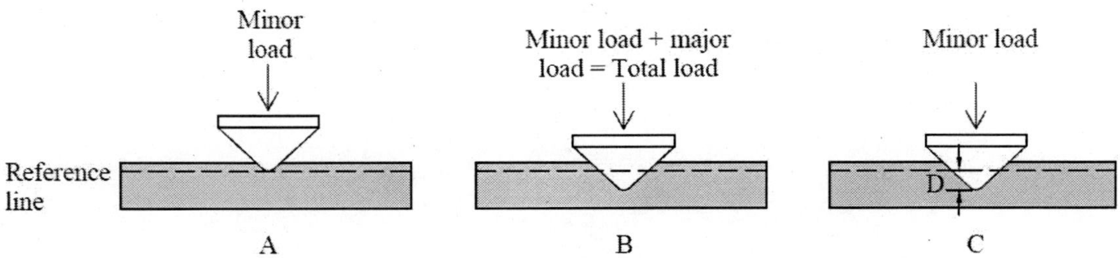

Fig. 3.24: Rockwell indentation

The following are the equations to determine hardness of a specimen using Rockwell hardness method, for conical and ball type indenters respectively.

For conical indenter,

$$\text{Rockwell Hardness} = 100 - \frac{\text{Indentation caused by Major load } (D)}{0.002} \tag{3.18}$$

For ball shaped indenter,

$$\text{Rockwell Hardness} = 130 - \frac{\text{Indentation caused by Major load } (D)}{0.002} \tag{3.19}$$

3.10.2 Internal friction

Internal friction is also known as mechanical spectroscopy, it can be used to analyse the atomic motions in solids brought on by applied oscillatory strain and heat activation. Internal friction is the ability of a vibrating material to transform its mechanical vibrational energy into heat, even when it is so well isolated that energy losses to its surroundings are minimal. The reduction of vibration amplitude or damping of a freely vibrating body, is the example of an internal friction. The measurement of internal friction has proven to be a very useful tool in recent years for learning more about the behaviour of gas atoms and other interstitially dispersed atoms in metals. Diffusivities, concentrations, terminal solubilities, surface reactions, precipitation events, mutual contacts, and interactions between interstitials and other lattice flaws can all be obtained from it. The process for measuring internal friction is straightforward in theory: after the metal sample vibrates, the amount of energy lost due to internal causes—a process that turns vibration energy into heat—during the oscillation is either directly or indirectly recorded. Internal friction can be caused by hysteresis, relaxation, or both, but in rare circumstances, resonance may also be a factor. Internal friction is attributed in the atomic model to relative displacements of lattice flaws caused by the vibration. In addition to interstitial atoms, other flaws that can cause internal friction include conduction electrons, phonons, vacancies, substitutional atoms, dislocations, grain boundaries, and domain walls.

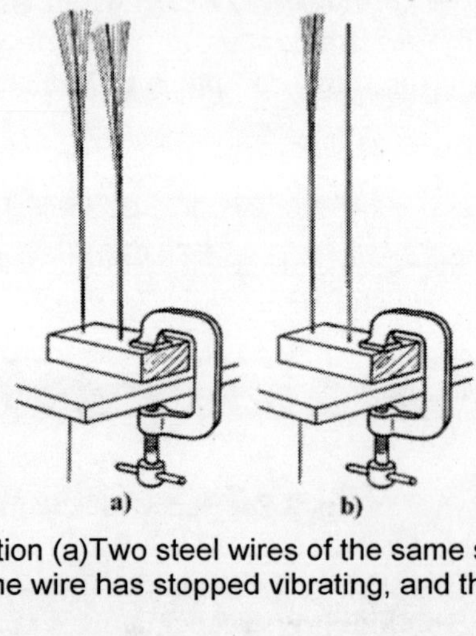

Fig. 3.25: Example of internal friction (a)Two steel wires of the same size begin oscillating at the same time (b) A litter later, one wire has stopped vibrating, and the other is still oscillating

Internal friction in fluids can be detected by watching damped oscillations of different liquids, such as water, oil, or glycerine, in transparent containers. The presence of internal friction in elastic substances can also be inferred from similar measurements of the various damping periods of oscillations of two metal wires of the same size (see Fig. 3.25).

3.11 PHYSICAL PROPERTIES

Physical properties play an important role from application point of view. Physical properties like weight, dimension and water absorption all play an important role. Consider an example of a bricks, one of the commonly used construction materials in India. A major reason for its popularity is the ease of field application. To ensure this, the bricks need to be of standard sizes; uneven sizes will create unstable construction. Hence dimensions, a physical attribute, plays an important role in the field application of bricks. Similarly, water absorption and specific gravity of aggregates is important for the mix design of aggregates.

From application point of view following material properties are important:

Density: Density indicates the weight of the material against the volume occupied by it. In many applications, like during mix design, the absolute volume of material is considered; then the density is reported in terms of absolute density or specific gravity. And, in other applications, like soil, the voids and pores are also considered in the volume occupied; then the density is reported in terms of bulk density.

Specific gravity is determined by measuring the volume occupied by a given weight of the material in a non-reacting liquid. Le Chatelier flask, pycnometer and density jar are some of the equipment for measuring specific gravity. While bulk density is determined by measuring the weight of the material occupied in a vessel of known volume. Bulk density is typically measured against the degree of compaction of material.

Size/dimension: Size plays an important role to ensure the homogeneous application of materials. In case of materials like aggregates, the size is measured using sieve analysis. While in case of materials like bricks and woods the size is measured using any typica scale. For each application an allowable variation in standard size is defined, known as tolerance. The measured size/dimension should not exceed the specified tolerance.

Rheology or flow behaviour: Rheology defines the flow behaviour of any flowable engineer materials. Among the various civil engineering materials, cement, bitumen, mortar, concrete and asphalt are applied in flowable or wet state. Rheology of these materials is either measured using conventional tests like slump flow, where degree of flow is observed in specific conditions, or modern equipment like rheometer, where degree of flow is observed over a range of conditions.

Water absorption: Many civil engineering materials require water during their application. Absorption of water increases the required quantity and affects the distribution of water; hence it should be accounted for during the application. For example, during brick-and-mortar construction, high water absorption of bricks will absorb water from the mortar and reduce the water available for cement hydration. Water absorptions is determined by measuring the percentage change in weight of a material in saturated surface dry (SSD) condition and oven dry condition.

Permeability and porosity: Porosity defines the voids, cavities and micro-cracks present in any engineering material. Porosity is either determined by measuring the amount of water absorbed by the material or by correlating bulk density and specific gravity.

When, pores inside an engineering material are interconnected, they allow for the flow of liquids and make the material permeable. Permeability allows for the transportation of fluids, which is a desirable property in some cases, like soil. In other cases, like concrete, fluids can carry harmful compounds like sulphates and carbonates which can damage the material and hence permeability is undesirable.

Thermal properties: Thermal properties like thermal expansion, thermal conductivity, softening point, melting point, flash point, etc., affect the performance of material during temperature changes. The required thermal property is important depending on the material and desired application. For example, thermal expansion is important to prevent dimension changes in steel and concrete structures. Thermal conductivity is important to lower the heat transfer in cases of walls. Similarly, softening point is important in bitumen to ensure that bitumen is flowable when heated and not at operational temperature.

Other physical properties also exist like colour, lustre, opacity, electrical conductivity, etc. Each of the physical property can be determined for all civil engineering materials. However, all physical properties are not relevant from application point of view, and hence should be carefully considered.

Additionally chemical and mineralogical tests also exist for civil engineering materials. These tests are important from material science point of view, when identifying the material. Therefore, chemical and mineralogical characteristics hold less relevance from application point of view.

UNIT SUMMARY

Mechanical characteristics and behaviours

- Strength – tensile, compressive, bending, torsional, etc.
- Behaviours – ductility, malleability, brittleness, resilience, etc.
- Defined properties – elasticity, proof strain, etc.

Different types of tests

- Tensile, hardness, bending, torsion, creep, internal friction, fatigue, fracture

Deformations and failure modes

- Elastic, plastic, creep, fatigue.

Fracture mechanics

- Toughness and energy absorption.
- Ductile-brittle transition temperature.

Physical properties

- Density, size, rheology, etc.

EXERCISES

Multiple Choice Questions

3.1 Which of the following properties is not associated with the deformation of a material:

(a) Toughness (b) Hardness

(c) Ductility (d) Resilience

3.2 A material deforming without change in applied loads can be associated with:

(a) Fatigue (b) Malleability

(c) Creep (d) Elasticity

3.3 Elasticity is defined as the region where material shows:

(a) Linear relationship between stress and strain, and deformations are reversible

(b) Linear relationship between stress and strain, even if deformations are irreversible

(c) Reversible deformations, independent of relationship between stress and strain

(d) None of the above

3.4 The relationship between lateral and longitudinal strain is defines as:

(a) Poisson's ratio (b) Young's modulus

(c) Bulk modulus (d) Rigidity modulus

3.5 Plastic deformation through twinning occurs in:

(a) Single crystal metals with BCC lattice

(b) Any single crystal metal, independent of lattice

(c) Any polycrystalline metal

(d) Any metal, either single or polycrystalline

3.6 Which of the following is true for a brittle material:

(a) Plastic deformations are absent

(b) Proof stress is determined using 0.2% offset strain value

(c) Failure is sudden

(d) All of the above

3.7 Hardness can be measured using

(a) Brinell testing machine (b) Charpy testing apparatus

(c) Troptometer (d) Universal testing machine

3.8 Ceramic performs excellent in terms of

(a) Electrical conductivity

(b) Thermal conductivity

(c) Transverse strength

(d) Compressive strength

3.9 Steel will undergo brittle failure if

(a) Transition temperature is above operational temperature

(b) Transition temperature is below operational temperature

(c) Steel has been exposed to nuclear radiations

(d) None of the above, as steel is a ductile material

3.10 Which of the following is true for fracture mechanics

(I) Failure depends on failure strength of material

(II) Failure depends on toughness of material

(III) Failure originates at one point and spreads though out the material

(IV) Failure originates at multiple locations causing a simultaneous failure

(a) I and III (b) II and III (c) I and IV (d) II and IV

Answers of Multiple Choice Questions

Answers of Multiple Choice Questions
3.1 (b), 3.2 (c), 3.3 (c), 3.4 (a), 3.5 (b), 3.6 (d), 3.7 (a), 3.8 (d), 3.9 (a), 3.10 (b)

Short and Long Answer Type Questions

3.1 Write a short note on elasticity of materials.

3.2 List the various properties affected by the plastic deformations in a material and write about them.

3.3 Explain the concept of plastic deformation in metals.

3.4 Write a short note on creep.

3.5 What are the advantages of ceramic over metals. Write a short note.

3.6 Explain ductile-brittle transition of steel.

3.7 Explain the concept of fracture mechanics, as compared to strength-based failures.

Numerical Problems

3.1 In a torsional test a cylindrical sample, length 0.3 m and diameter 0.1 m, is subjected to rotation of 0.001 radians using a torque of 10 kN. Calculate the modulus of rigidity and shear stress at the extreme fibre of the cylinder.

3.2 In Brinell hardness test, a 10 mm ball applied through a pressure of 1 kN load resulted in an indentation of diameter 2 mm. Determine the Brinell hardness number.

3.3 Calculate the critical resolved shear stress, if for an FCC single crystal, slip plane and slip direction are given by the vector $[1\ 1\ 1]$ and $[1\ \bar{1}\ 0]$, respectively. Consider the acting tensile stress as 5 MPa acting along the X-direction $[1\ 0\ 0]$.

PRACTICAL

Relevant practical for this unit have been included within the unit itself.

KNOW MORE

Non-Destructive Testing

Most testing methods require destruction of samples and suitable for laboratory scale testing. However, during the field application destructive testing of structures is not possible. This challenge is resolved by non-destructive tests (NDT). NDT uses equipment to measure the physical attributes and mechanical properties of a material in non-destructive (typically elastic) zone to correlate the actual behaviour of material.

The most commonly used NDT methods in the field of civil engineering are based on:

Ultrasonic: Ultrasonic waves are a type of volumetric testing methods. Ultrasonic waves can be used to identify the structural integrity of a material, i.e., lack of porosity and microcracks. As structural integrity indicates the lack of defects in a material, ultrasonic tests can be used to assess the mechanical characteristics of a material with similar level of structural integrity. Ultrasonic pulse velocity (UPV) is one of the most widely used NDT in the field of civil engineering.

Liquid penetrant: In these types of tests, a special type of liquid penetrant which can be easily detected is applied over a surface. After the application the presence of liquid penetrant is used for describing the surface level structural integrity of materials. This type of NDT is commonly used for testing of joints and welds.

Magnetic: Ferric metals respond to magnetic fields and generate distortion in magnetic fields. Monitoring of distortion in magnetic field can be used for identifying changes in ferric material. This type of NDT is widely used for locating reinforcements and assessing their degree of corrosion in RCC structures.

Radiographic: Radiographic NDT tests are similar to NDT, but can be used for producing exact mapping of internal structure of materials. These types of tests are expensive and hence have limited field applications. However, they have significant application in improving the understanding of material fracture. This includes test like CT scan, X-ray and SEM.

Visual: Visual tests are often employed to assess the mechanical strength of material. However, these tests are highly subjective and only provides a qualitative assessment. Visual NDTs can range from visible crack detection to colour change in steel from rusting.

Mechanical: Mechanical NDTs are focused on limiting the testing region to reversible deformations in a material. Mechanical NDTs should be carefully conducted to ensure that residual strains are not left in the material. Rebound hammers and nano-indenters are some of the commonly used mechanical NDTs.

In addition to the above tests several new sensing methods are being developed to identify the structural integrity of a material, like thermal imaging. NDTs are typically less accurate but present an approximate evaluation of materials. Therefore, often more than one NDT is applied to justify their safe engineering application of materials.

REFERENCES AND SUGGESTED READINGS

1. K. Komvopoulos, *Mechanical Testing of Engineering Materials*, Cognella, 2017
2. M. R. Islam, *Civil Engineering materials – Introduction and laboratory testing,* CRC Press, 2020

3. J. Roesler, H. Harders, M. Baeker, *Mechanical Behaviour of Engineering Materials*, Springer, 2007
4. N. E. Dowling, S. L. Kampe, M. Kral, *Mechanical Behaviour of Materials*, Pearson, 2019
5. V. Raghavan, *Materials Science and Engineering*, PHI Learning Pvt. Ltd., 2021
6. N. Subramanian, *Building Materials Testing and Sustainability*, Oxford University Press, 2019.
7. J. P. Patel and G. H. Upadhyay, *Material Science*, Atul Prakashan, 2001.

Dynamic QR Code for Further Reading

4 Standard Testing & Evaluation Procedures

UNIT SPECIFICS

Through this unit, we have discussed the following aspects:

- *Various standard testing and evaluation procedures;*
- *Different types of mechanical tests;*
- *Preparation of sample and procedures for various tests;*
- *Variation in mechanical tests for different application requirement;*
- *Interpretation of test results for identifying mechanical behaviours*

The different types of tests and evaluation procedures for civil engineering materials are discussed to create an understanding for the material testing and develop necessary knowledge for the interpretation of test results.

Along with a large number of multiple-choice questions and questions with short and long answers, the unit also includes a list of references and suggested readings that one can use as practice and grow their knowledge on material testing.

RATIONALE

This fundamental unit on standard testing & evaluation procedures helps the reader to familiarize with the different mechanical and non-mechanical test of civil engineering materials. The unit describes the various tests associated with strength, impact, fracture toughness, fatigue and creep of the material. The unit also describes the working principle of mechanical tests, preparation of test samples, modifications is test and interpretation of test results. The discussion is followed by some problems related to material testing, which will further help in developing a clear idea of the concerned topics on the subject.

A good grasp of testing and evaluation procedures along with understanding of their results is necessary for selecting of suitable materials for construction of buildings and other structures. Understanding the mechanical behaviour of different materials from their corresponding tests, will help in safe engineering application of the said material.

PRE-REQUISITES

Physics: Mechanics (Class XII).

Chemistry: Lattice structures (Class XII).

UNIT OUTCOMES

As outcome of this unit, one should be able to:

U4-O1: Understand mechanical tests and their working principles

U4-O2: Prepare test samples and test setups as per standard testing procedures

U4-O3: Modify the tests as per requirements of the application for a given material

U4-O4: Record the results of the tests

U4-O5: Interpret the test results to identify the characteristics of engineering materials

Unit-4 Outcomes	EXPECTED MAPPING WITH COURSE OUTCOMES (1- Weak Correlation; 2- Medium correlation; 3- Strong Correlation)							
	CO-1	CO-2	CO-3	CO-4	CO-5	CO-6	CO-7	CO-8
U4-O1	-	-	2	1	1	1	2	2
U4-O2	3	2	3	3	-	1	2	1
U4-O3	2	2	2	2	-	1	1	1
U4-O4	-	3	2	2	3	2	3	3
U4-O5	-	-	-	-	3	3	3	3

4.1 MECHANICAL TESTING

Mechanical characteristics are one of the fundamental attributes of any civil engineering material. Materials may be designed for load bearing purposes, like columns where they have to carry the loads without failure, or even non-load bearing purposes, like partition walls where they have to carry their self-weight only. Even in case on applications like paint, scratch resistance, strip resistance and hardness become important mechanical characteristics. Therefore, irrespective of the potential application, mechanical characteristics play an important part in the application of civil engineering materials. These mechanical tests include strength test(s), impact test, fatigue test, among several others.

4.1.1 Laboratory for mechanical testing

Mechanical testing involves multiple different testing protocols depending on the nature of applied load and support conditions of the specimen. These test protocols are used to effectively create the various different forms of scenarios which the material may encounter during the field application. There exists a large number of laboratory equipment for conducting the tests based on multiple factors like type of material, size of testing specimen, property to be tested, standards and guidelines of a given country and operational complexities. These equipment's, found in a typical laboratory, can be grouped into following categories;

Laboratory equipment for strength testing:

Strength tests are conducted to study the performance of an engineering material under gradually acting loads. The major attribute for analysis is studying the various deformations and resulting load resistances. The test setup includes three core components, i.e., support condition, loading mechanism and deformation measurement devices.

The support conditions are used to hold the specimen and provide a balancing force against the applied loading. Different support conditions can be used to generate different strength tests like compression, tension, torsion, split tensile and flexural. Among the several equipment, universal testing machine (UTM) is most commonly used in laboratories (refer Fig. 3.10). UTM can generate a large number of support conditions and perform a large number of strength tests. The loading direction in an UTM remains the same, typically from bottom to up, while the different tests are generated by adjusting the specimen support conditions. Fig. 4.1 shows the different support conditions for a cylindrical specimen under an UTM.

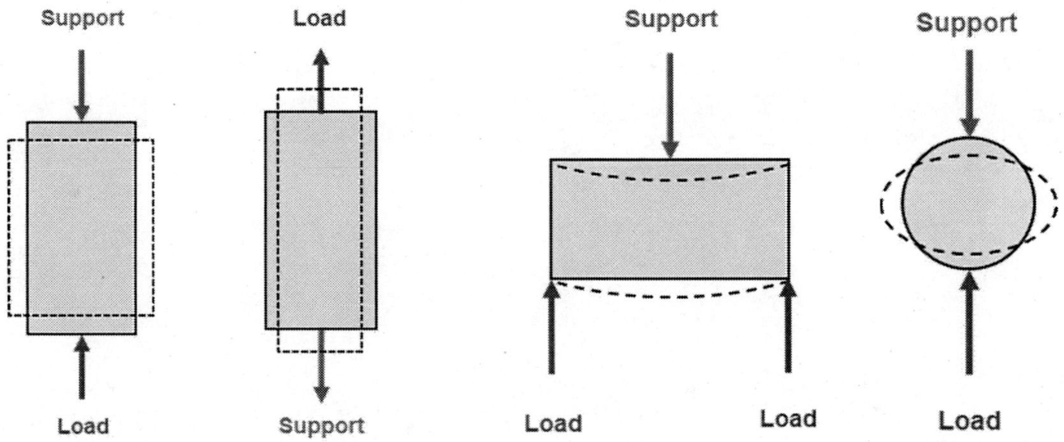

(i) Compression test (ii) Tension test (iii) Flexural strength test (iv) Split tensile test

Fig. 4.1: Support conditions in an UTM

Loading mechanism usually involves hydraulic jack and piston to apply the desired force. The loading mechanism is typically controlled by the displacement of the piston. However, several different loading mechanisms also exists which include water pressure, direct loads and motors. Furthermore, loads can also be applied from multiple directions to create bi-axial and tri-axial loading. Commonly uniaxial loading is adopted for most engineering materials. While, tri-axial loading is commonly used for clay specimen. In addition to the loading mechanism a load cell is attached along the axis of loading to measure the value of applied load. The load cell is commonly attached to the supports, i.e., on top of the specimen.

Deformation measurement devices typically consists of either displacement measurement devices, like LVDT, or strain measurement devices, like strain gauges. The value of displacement is essential for determining several mechanical characteristics including elasticity, yield strength, Poisson's ratio, etc. Similar, to loading mechanism the deformations can be measured along different axis. Since data gathering is required on a continuous basis, the load cell and strain gauge/LVDT are often connected to data logger for recording data.

Laboratory equipment for impact test:

Impact test involves application of load in the form of impulse in a specified short duration of time. The equipment consists of two essential factors, i.e., load and a load launching mechanism. A third component of deformation measurement is also present in some equipment. Similar to strength test, impact test can also have different support conditions, but mostly flexural support condition is used. Charpy test and Izod test are two most commonly used test setups for measuring materials toughness. While drop impact test is commonly used for measuring repeating effect of impact loads below fracture energy. In impact test the load is typically dropped from a specified height under the action of the gravity. The load is then allowed to strike the sample and the energy absorbed by the specimen after failure is measured, as its toughness. Fig. 4.2 represents the two form of impact test setups.

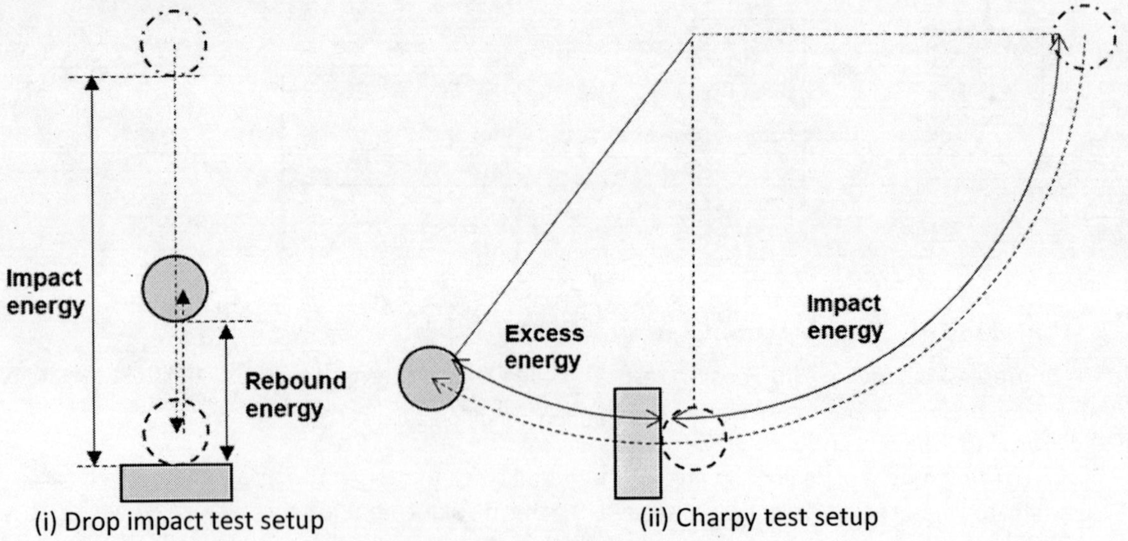

Fig. 4.2: Impact test setup

Laboratory equipment for fatigue test:

Fatigue test involves measuring the effect of repeated loading cycles, over a long period of time. The test setup is similar to strength test, except for the loading mechanism. In loading mechanism, the loading and unloading rate along with load value are controlled to apply a cycling load. Similar to strength test, fatigue test can also have different support conditions.

Laboratory equipment for creep test:

Creep test is also a derivative of strength test equipment. In this test all-essential components are same as strength test; the only variation is the loading rate. In creep test a constant value of load is applied while the deformation is measured over a very long period of time. Similar to strength test, creep test can also have different support conditions.

These are basis test setups used for assessing the failure related mechanical characteristics of engineering materials. Other mechanical characteristics like hardness, scratch resistance and abrasion resistance do not cause mechanical failure of material during civil engineering application. Thus, their discussion is not much extended in this chapter.

4.1.2 Discussion about mechanical testing

Mechanical tests are often subjective to the test conditions in which they are performed. For example, in compression test a higher loading rate will result in higher compressive strength and vice versa. In order to prevent discrepancies, mechanical tests are conducted using a specific set of instructions. In India, Bureau of Indian Standards (BIS) prepares the guidelines for the preparation and testing of samples. Following section presents some of the commonly adopted test protocols applied form mechanical testing of widely used civil engineering materials:

Determining the properties of steel under tension

Steel is commonly used as tensile force bearing member. Material testing is primarily done to create the proper stress-strain diagram in order to predict how steel will behave. For determining the behaviour of steel or similar materials under tension, IS 1608 suggests the following steps:

- Take the given sample and determine its cross-sectional area (S_o).
- Use the cross-sectional area to calculate the gauge length for the test specimen, using the equation $L_0 = 5.65\sqrt{S_0}$. Gauge length will be used for determining the elongation at any point throughout the test.
- Mark the gauge length in the form of straight along the longitudinal axis of the specimen. Here, the longitudinal axis represents the direction along which tensile strength is to be determined. Make sure the gauge length is centrally located and provides unmarked space on either end for holding the specimen.
- Fix the unmarked ends of the test specimen using the griping mechanism provided in a universal testing machine, or a similar test apparatus.
- Operate the machine to separate the two ends of the test specimen, at rate specified in IS 1608 based on the type and desired property of the specimen.
- Take a note of elongation of the specimen, measured as the change in gauge length and load resistance offered by the specimen. Perform the test till the failure of the specimen.
- The stress (σ) is calculated as the load resistance offered by the specimen per unit cross-sectional area, and strain (ϵ) is calculated as the change in length per unit gauge length of the specimen.
- The values of stress at different strains are plotted to understand the various properties of test material.

It should be noted that the gauge length and grip length are based on the type of specimen being used. The method describes the sample preparation for steel members with symmetrical cross-section areas having diameter more than 4 mm, and sheet sections having thickness greater than 3 mm. For other dimensions or non-symmetrical cross-sections different gauge length and grip length shall be adopted as per IS 1608. Fig. 4.5 shows the different types of cross-sections where gauge length is determined using Eq. 4.1.

$$L_o = k\sqrt{S_0} \qquad \text{where} \qquad k = 5.65 \tag{4.1}$$

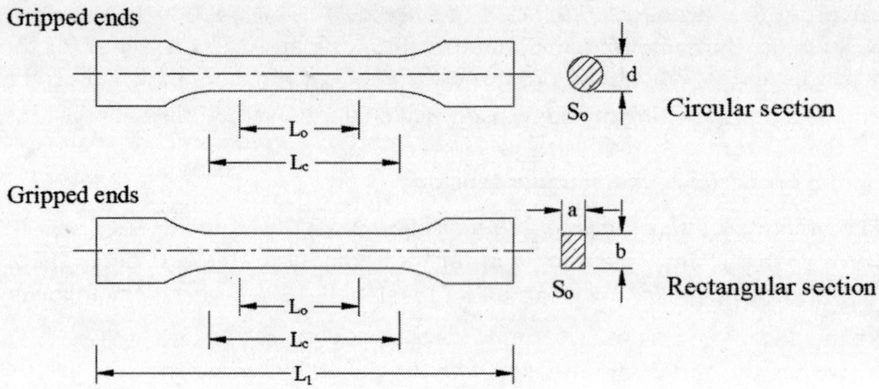

S_o	: Original cross-sectional area
L_1	: Total length
L_o	: Gauge length
L_c	: Length of test specimen
a, b	: Width and thickness of rectangular section
d	: Diameter of bar

Fig. 4.3: Test Pieces of Circular and Rectangular Cross-section

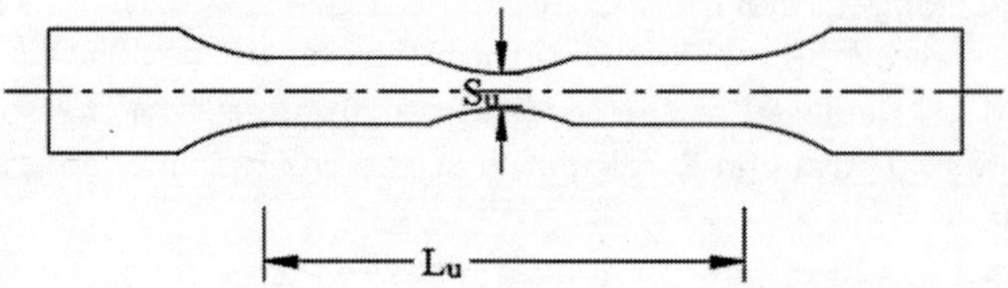

S_u	: Minimum cross-sectional area after fracture
L_u	: Final gauge length after fracture

Fig. 4.4: Test pieces of circular cross-section after fracture

Fig. 4.5: Steel samples for tension tests

Determining the properties of concrete under compression

Concrete is commonly used for bearing compressive forces in engineering applications. Unlike steel, concrete fails as a brittle material and hence stress strain curves are of less importance than ultimate compressive strength at the time of failure. The method for determining compressive strength of concrete is as per IS 516 and can be described as follows;

- Proportion the ingredients of concrete as per IS 10262, or similar relevant standard.
- Mix the ingredients to achieve a consistent wet concrete.
- Test the workability of wet concrete. If the mix fails to achieve the desired workability, then adjust the proportion and repeat the steps.
- Cast the concrete in a cubical mould of size 15 × 15 × 15 cm. Other moulds may be used as per the test and adopted standard.
- Compact the concrete and keep it under humid conditions without disturbance.
- Demould the concrete after 24 h and cure the test specimen. Capping may be performed if surfaces are not parallel.
- At the desired age of testing, concrete specimen is cleaned with wet cloth and weighed.
- Place the sample under compression set up and apply the loading rate of 14 N/mm^2/mm.
- Perform the test until the failure of specimen.

4.1.3 Naming system for various ferrous and nonferrous metals

Metals for engineering applications are often identified by their commercial nomenclature. It is important to understand the naming system to select a suitable metal for testing. As per IS codes there are two types of naming systems used for both ferrous and nonferrous metals. The first naming system is based on mechanical characteristics of the metals and is used for steel and iron. The second naming system is based on chemical composition and is used for all metals. For non-ferrous alloys the mechanical characteristics are usually designated corresponding to their chemical composition.

Naming system is based on mechanical characteristics

The naming is based on a five-part system based on IS 1762. The first part represents whether ultimate strength (Fe) or yield strength (FeE) of steel is used for naming system. The second part represents the minimum value of strength in terms of N/mm^2. Third part represents chemical symbols for elements present in the steel. The fourth part provides information on special characteristics of steel. And the final part represents the application for the steel. It should be noted that it is not mandatory to use the last three parts in the naming system. Fig. 4.6 shows a representation of the naming system. In the fourth part the special characteristics are shown in Table 4.1.

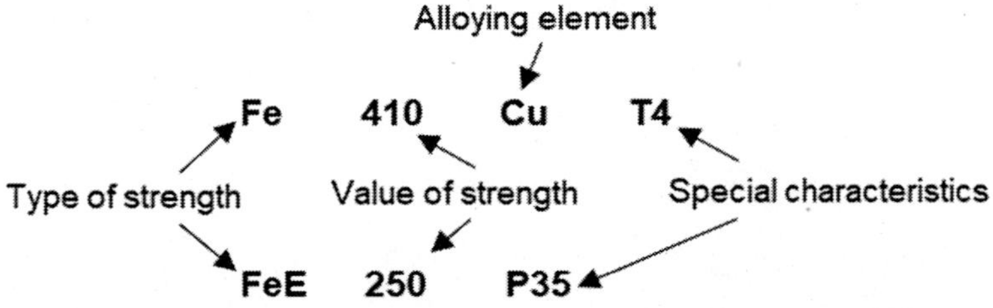

Fig. 4.6: Naming system for steel based on mechanical characteristics

Table 4.1: Special characteristics of steel reported in the naming system as per IS 1762

Symbol	Representation
R, K	Method of oxidation (rimming steel, killed steel)
Q1	Quality of steel; number varies from 1 to 5
P25	Degree of purity in terms of maximum phosphorous content; number indicates 100 times the maximum percentage of phosphorous content rounded off to the nearest integer
SP44	Degree of purity in terms of maximum sulphur and phosphorous content; number indicates 100 times the maximum percentage of sulphur and phosphorous content rounded off to the nearest integer
W, W_1	Weldability guarantee (fusion weldable, resistance weldable)
B	Resistance to brittle fracture; Number is added after the letter to indicate fracture energy as per Charpy V-notch test
S1	Surface condition; number varies from 1 to 7
D1	Formability; number varies from 1 to 3
F1	Surface finish; number varies from 1 to 14
T1	Treatment; number varies from 1 to 14
H	Letter H is used to denote elevated temperature properties
L	Letter L is used to denote low temperature or cryogenic properties

Naming system is based on chemical composition

All metals are commonly identified by their chemical compositions. The nomenclature can either based on alphanumeric designation or numeric designation. Different naming systems are followed as per the type of metal as shown in Table 4.2. Since the naming system based on chemical characteristics is less relevant from testing point of view, the same has not been discussed in detail.

Table 4.2: Various naming systems based on chemical compositions

Metal and their alloys	Nomenclature	Relevant standard
Iron/steel	In case of steel the naming system denotes the major chemical constituents of the metal. The naming system slightly differs for different types of steel but is loosely based on a four-part system which indicates type of steel, carbon content, alloying elements and special characteristics respectively. For example, T75Cr4 indicates unalloyed tool steel containing about 0.75% carbon and 1% chromium. It should be noted that numbers have a different multiplier for different elements.	IS 1762
Ferroalloys	In ferro alloys, the naming system is used to denote chemical constituents and their composition in a four-part system which indicates carbon, alloy, major constituent and phosphorous, respectively. The naming system for ferro alloys doe not use the symbol of carbon, awhile rest have their usual notations. For example, 2MnSi18P6 indicates about 2% Carbon, 18% Silicon, 6% Phosphorous in Manganese as base alloy.	IS 2085
Aluminium	Follows either a 'five-digit number system' for wrought aluminium and its alloys, and 'four-digit number system' for casting alloys, castings and primary ingots. In both systems first digit indicates major alloying element and second digit indicates the rounded off mean value of alloying element. The third, fourth, and fifth digit indicate minor elements in decreasing order. For example, 31000 indicates wrought aluminium alloy with manganese as alloying metal at about 1%.	IS 6051
Copper	Follows a two-part naming system indicating chemical composition and special characteristics. All symbols should be followed by the chemical symbol for copper, i.e., Cu. For example, Cu Zn 35 Sn 2 indicates copper alloy with about 35% zinc and 2% tin.	IS 2378

4.1.4 Preparation of test specimen

Most engineering materials are supplied in the form they are to be used in the field, i.e., no material modifications are made. Thus, these materials are directly used for mechanical testing. The only preparations required is the shape modification to meet the specifications of the testing equipment. For example, during Charpy impact test of steel plates the sample will be machined so that the test specimen has the required V-notch. On the other hand, several engineering materials are prepared on site during their application itself. These materials require preparation before testing. For example, the test specimen for compressive strength of cement is prepared at a specified mixing water determined during preparation of specimen. Table 4.3 provides a list of various engineering materials which require preparation along with relevant standards.

Table 4.3: Relevant standards for material preparation

Material	About material preparation	Relevant standard
Cement	Done using the standard consistency of cement. • First water required standard consistency cement is determined. • This is used to calculate the amount of water required for mixing with a given amount of cement and sand. • After proportioning the mix, mortar is prepared, cast and cured before final testing.	IS 4031
Mortar	Done using working consistency and desired grade of mortar. • First a mix proportion of cement, lime, pozzolana and sand is selected as per the desired grade/strength of mortar. • Then water is added and mixed to prepare the mortar. • Water is adjusted till the desired working consistency is achieved. • Mortar is then cast and cured before final testing.	IS 2250
Concrete	Done using the mix proportioning guidelines. • First the specific gravity, water absorption and particle size distribution of different components is determined. • A value of cement content and w/c ratio is selected. Value of other additives like admixtures and fibres is also selected. • Based on specific gravity and bulk density of concrete, weight ratio of different constituents is calculated. • Aggregates are proportioned in fine and coarse aggregates. • Water correction factors are applied as per the water absorption, and type of aggregates. • Proportioned constituents are mixed and tested for workability. • The selected values are adjusted till desired workability is achieved. • Concrete is then cast and cured for final testing.	IS 10262
Aggregate	Done for determining the bulk properties of aggregates. • Aggregates are surface dried. • Sieving is done to obtain the aggregates in specified size range. • Aggregates are placed in their specified mould and tested without demoulding.	IS 2386 (Part IV)

<table>
<tr><td>Bricks</td><td>Done for obtaining smooth parallel surfaces<ul><li>Any unevenness on the selected parallel surfaces is removed.</li><li>Sample is submerged in water for 24 h. After removing surplus surface water is drained.</li><li>Frog is filled and face flushed with mortar.</li><li>Mortar is cured for 24 h and then cured under damp cloth for 24 h followed water immersion for 3 days.</li><li>Sample is then cleaned from surface water before final testing.</li></ul></td><td>IS 3495</td></tr>
<tr><td>Asphaltic concrete</td><td>Done as per the requirement is Marshall method of design<ul><li>First a suitable grade of bitumen, and aggregate gradation is selected. Bitumen content and fillers are also selected.</li><li>Materials are heated and mixed to prepare a consistent mix.</li><li>The mix is placed in a preheated mould and compacted with the rammer for a fixed number of blows.</li><li>The values of Marshall stability and other relevant properties are determined.</li><li>Proportion is adjusted till the required values are achieved.</li><li>The material can now be placed, cooled and used for final testing.</li></ul></td><td>IRC 29</td></tr>
</table>

4.2 STRENGTH TESTS

The primary goal of the strength tests is to determine the mechanical strengths and deformation behaviour of engineering materials. The specific test method differs for different materials and tests, and various standards are used. Among these standards two of the most commonly used standards are IS 516 for mechanical characteristics of concrete and IS 1608 for tensile testing of metals. After the tests have been performed the results are obtained in terms of load vs. deformation or stress vs strain. The graphs are then interpreted for determining the mechanical characteristics of the given material. It should be noted that strength tests can be performed for different loading conditions and hence stress-strain graphs can be prepared for compression, tension, torsion, shear or flexure. The typical stress-strain curves possible after the strength tests are shown in Fig. 4.7.

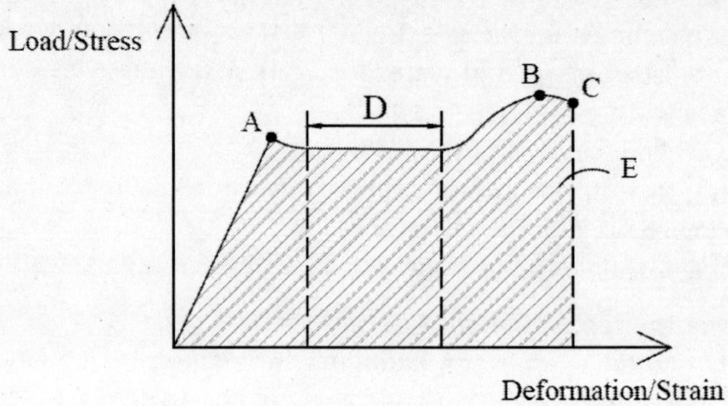

(a) Ductile material

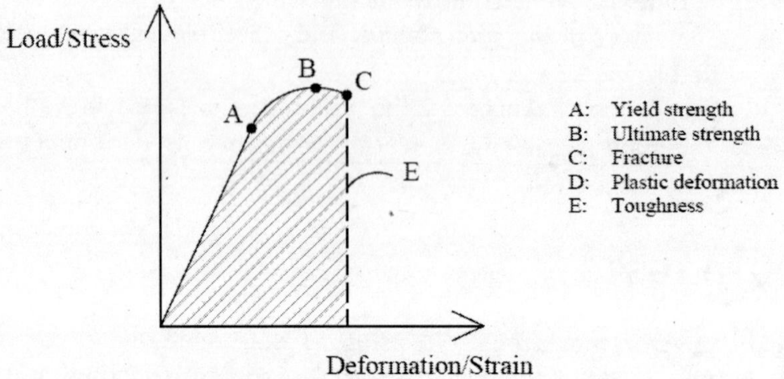

(b) Brittle material

Fig. 4.7: Typical stress-strain curves under tension

It can be observed from Fig. 4.7 that stress strain curve can vary for different materials, on the basis of deformation patterns. Each engineering material in exhibits two forms of deformation, i.e., elastic deformation and plastic deformations. The stress-strain curves can be interpreted to understand these deformations and the mechanical characteristics of different engineering materials.

4.2.1 Elastic deformation

Elastic deformation refers to the region where the material regains its shape after the applied loading has been removed. The maximum stress resisted by the body up to which elastic deformation occurs is known as yield strength and the point in stress-strain curve is known as yield point. Yield strength in case of ductile material is evident by the sudden change in stress-strain curve where material starts to deform without any additional loading. However, in case of brittle material there is no definite change in deformation behaviour. And thus, to determine the yield stress, a line parallel to initial tangent modulus of elasticity is potted at an offset strain of 0.2%. The point of intersection of the imaginary line and stress strain curve is known as yield point and corresponding stress is known as proof stress (Fig. 4.8).

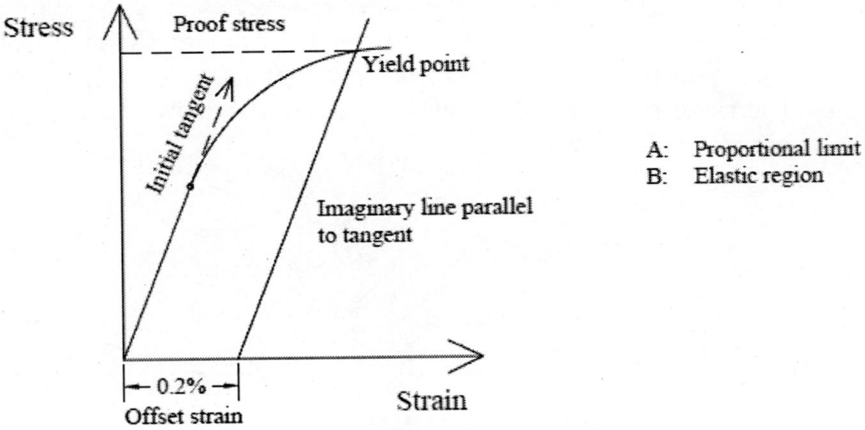

Fig. 4.8: Yield point in brittle material

The elastic region also defines modulus of elasticity and resilience of the material. The modulus of elasticity is defined as the slope of stress-strain curve. At low stress values, the stress strain curve shows a linear correlation between stress and strain. This region of linear correlation is defined as the proportional limit. Hooke's law is valid in the proportionality limit and is mostly used for calculating the modulus of elasticity. Beyond proportionality limit there is a non-linear relationship between stress and strain. In brittle material the proportional limit may be very small than the yield point, and hence different slopes are used to define the various modulus of elasticity, like, initial modulus of elasticity, secant modulus of elasticity and tangential modulus of elasticity. Resilience of the material can be calculated using the area of the stress-strain curve, up to the yield point.

4.2.2 Plastic deformation

After yield point, ductile materials, like metals, undergo deformation without any additional load resistance. These deformations are known as plastic deformations. Plastic deformation occurs when applied load exceeds yield stress. Quantitatively, plastic zone is used to identify ductility and toughness of the material. On the other hand, qualitatively, zone of plastic deformation, helps in comparing the properties like malleability.

4.3 IMPACT TEST AND TRANSITION TEMPERATURES

4.3.1 Impact test

Impact test are used for determining the fracture energy for a given material. In this test a sample of with a notch is placed between the supports. A hammer is then dropped such that the hammer strikes at the back of the notch. The notch ensures that the sample fails at the specified cross-section. Energy absorbed by the sample till its failure is known as fracture energy. Fig. 4.9 shows the sample under impact test.

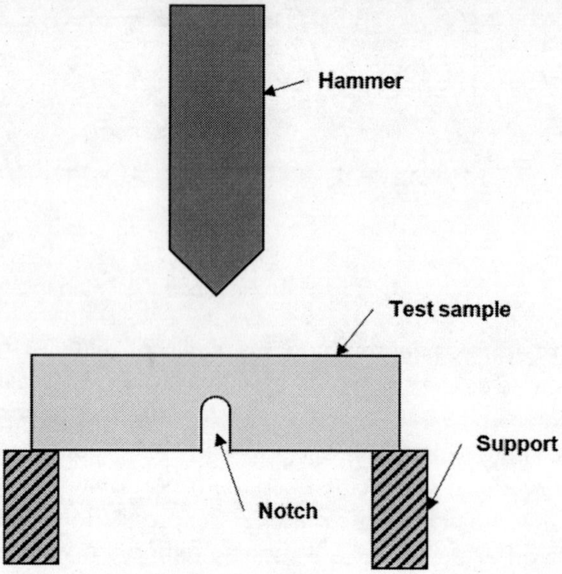

Fig. 4.9: Impact test

There are three types of impact tests listed in IS codes for metals, i.e., Charpy impact test (IS 1499; IS 1757), Izod impact test (IS 1598) and beam unnotched impact test (IS 5070). For other engineering materials a similar variant of the test is adopted. On the basis of the test the size of specimen and impact energy are specified in the respective standards. Once the sample is placed between the support grips the hammer is dropped to create and impact. The initial impact energy of the hammer and residual energy in the energy in the hammer after fracture are observed to calculate the impact energy using Eq. 4.2. The impact energy per unit volume is known as toughness and can also be determined using the area under the stress-strain curve.

$$\text{Impact energy} = \text{energy of hammer before impact} - \text{energy of hammer after impact} \qquad (4.2)$$

4.3.2 Transition temperature

As previously discussed in Section 3.12, that metals like steel change their behaviour from ductile to brittle when temperature falls below transition temperature. In order to determine the transition temperature of metals, the notched samples are tested using a pendulum-style impact tester. Energy absorbed during the fracture by conducting the tests at various temperatures. The results of tests are plotted in the form of an absorbed energy vs temperature graph, which theoretically takes the form of a S shaped curve, as show in Fig. 4.10. The transition curve looks more or less horizontal at higher temperatures and, is connected to the lower temperatures by a sharp line. The transition temperature range is represented by the vertical line, which in this case spans the range of -50°C to -40°C. The transition does, however, frequently occur across a much wider temperature range. As a result, it is now common practise to refer to the transition temperature as the level that occurs when the energy absorption drops to a specific minimum level, with 27 J being a value that is frequently used in European countries.

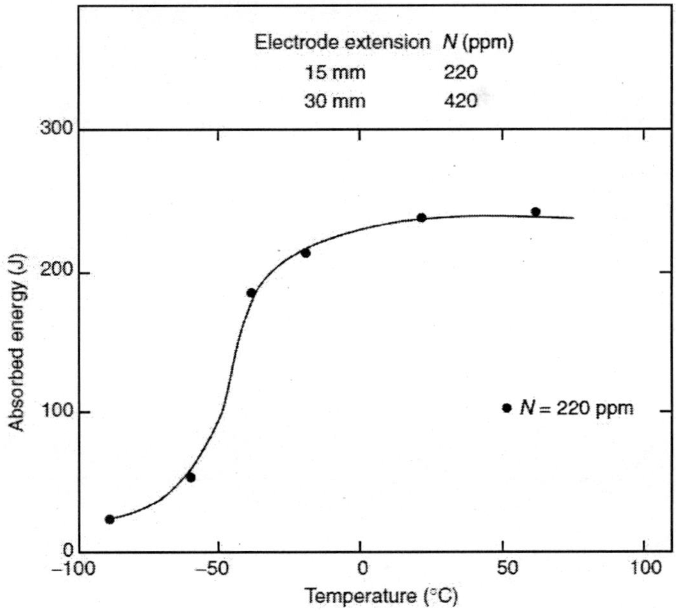

Fig. 4.10: Charpy V-notch curve for a 0.06C 0.5Ni steel weld metal
(Source: J. Lancaster, The technical background, Eng. Catastrophes. (2005) 139–189. https://doi.org/10.1533/9781845690816.139.)

At temperatures below the transition temperature, steel plates with an acute crack will fail brittlely if the crack is wide enough and the applied stress is strong enough. On the contrary, uncracked material, will behave typically ductile under pressure or when bent at such temperatures. Fig. 4.10 illustrates the properties of an ASTM A533 medium-tensile steel used to build thick-walled pressure vessels for nuclear power facilities. The fracture energy is plotted for the temperature range -300°F to +100°F (-184°C to +38°C). The upper diagram displays the fracture toughness and yield strength for different temperatures.

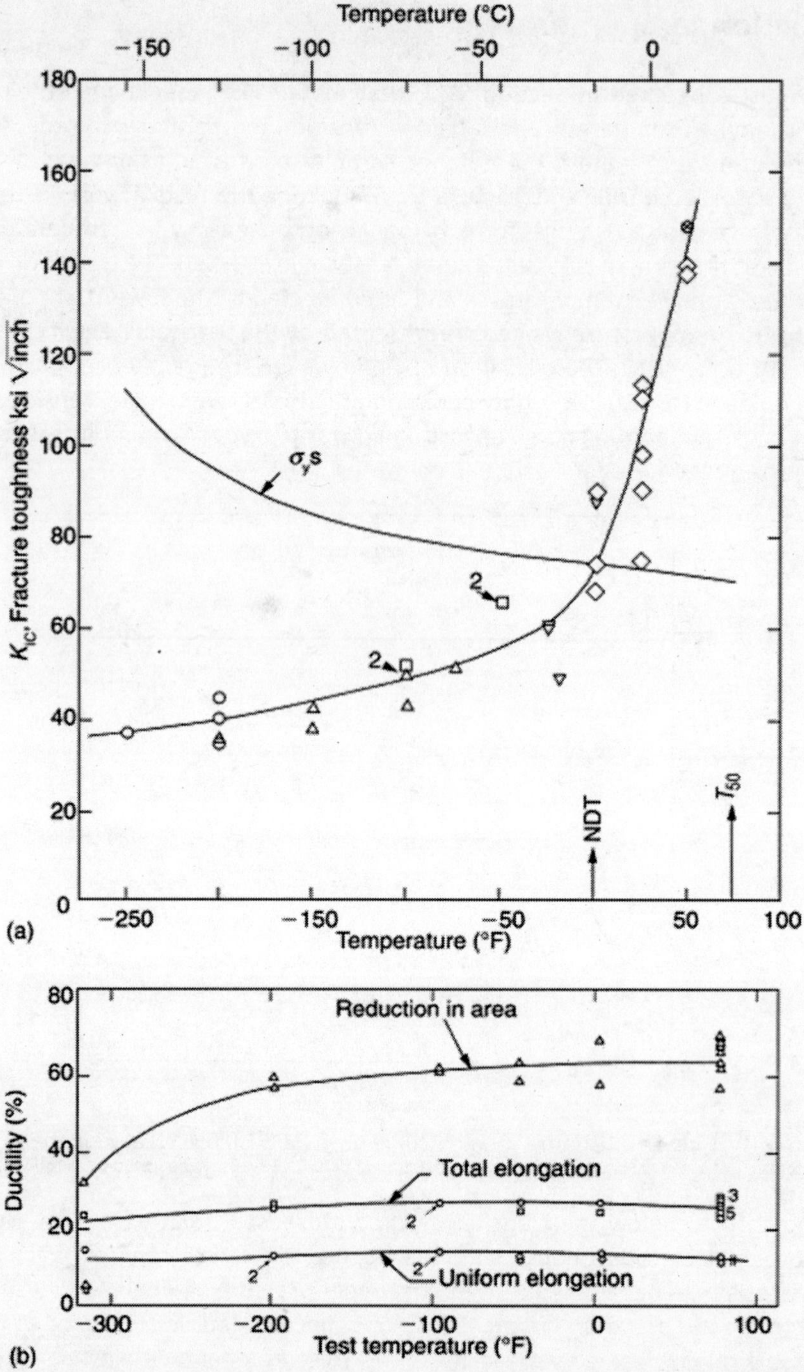

Fig. 4.11: Properties of a medium-tensile pressure vessel steel as a function of temperature: (a) fracture toughness and yield strength; (b) ductility as measured in a normal tensile test (after Wessel6).
(Source: J. Lancaster, The technical background, Eng. Catastrophes. (2005) 139–189. https://doi.org/10.1533/9781845690816.139.)

The ductility as determined by a typical tensile test is depicted in the lower diagram. Over the transition temperature range, there is barely any difference in elongation or loss of area, and values only begin to decrease below about -200°F. In other words, although steel can exhibit brittle behaviour at low temperatures, this behaviour is unusual and needs impact energy and existence of a sharp crack.

It goes without saying that lowering the ductile-brittle transition temperature below the temperature at which the structure must function reduces the danger of brittle failure. The grain size is the most crucial element in this case for low-carbon (mild) steel in the undamaged state. The transition temperature decreases with grain fineness. The rolling regimen to which the steel is subjected in manufacturing plates or sections, in turn, determines the grain size. Similarly adding elements like nickel to steel decreases the transition temperature. This schedule is frequently utilised when fabricating pressure vessels and pipework for process plants. For temperatures as low as -50°F, carbon steel with constrained impact characteristics is employed.

4.3.3 Embrittlement

Similar to transition temperature, embrittlement is another phenomenon which can cause loss of ductility. Strain cracking and temper embrittlement are two forms of embrittlement that have been linked to brittle fracture in steel. When steel bearing free (uncombined) nitrogen is strained and heated, either concurrently or afterwards, to a temperature in the neighbourhood of 200°C, strain ageing occurs. Nitrides begin to precipitate and cause embrittlement. In Germany and Belgium, several welded steel bridges were built in the 1930s. These were made using Bessemer converter steel, which could include free nitrogen due to the manufacturing process. Many of these bridges collapsed due to brittle fracture, and it was discovered that small cracks from welding had aged due to stress at the tip. These shallow, embrittled cracks are highly effective at starting unstable fractures. The solution was to mix in a little bit of aluminium with the steel. This refines the grain and boosts the yield strength while also removing the free nitrogen. The issue is unlikely to recur because Bessemer steel is only made in very tiny quantities these days.

Steel changes into the face-centred cubic structure known as austenite at high temperatures. The initial austenite grain boundaries are still present after cooling, but the material returns to its body-centred cubic form, ferrite. These areas have a disorganised lattice structure, and impurity atoms have a propensity to diffuse there. Such diffusion occurs during slow cooling starting at a temperature just about 600 °C and may result in embrittlement. The alloy composition, impurity content, and cooling rate all affect embrittlement to varying degrees. In the 1960s, brittle failures of a number of boiler drums were observed during hydrostatic testing. In these instances, copper, an alloying element that was purposefully added, served as the embrittling agent. The in-question alloy is no longer in use. Temper embrittlement appears to be the likely cause of the decrease in impact strength that results from the pressure vessels' post-weld heat treatment. This and other related issues should be reduced with increased steel purity.

4.4 FRACTURE TOUGHNESS TEST

4.4.1 Fracture mechanics – Background

Fracture mechanics deals with the mechanical behaviour of a cracked specimen. Fracture mechanics covers the load resistance, crack propagation and failure of material with cracks. During engineering applications, it is often possible that a material may be accidentally cracked during installation or operation. This crack propagates during operation and results in the failure of the structure. If the crack propagation is fast, then the material will fail abruptly and will cause major damage. On the contrary if the crack propagation is resisted by the material, then the cracks may be detected and preventive actions taken before the occurrence of failure. The resistance of materials to growth and propagation of crack is defined in terms of fracture toughness.

4.4.2 Fracture toughness – Different materials

Fracture toughness is a crucial factor in determining the reliability of engineering constructions. Fracture toughness is calculated as a function of applied load, crack geometry and specimen geometry. Fracture toughness is typically determined by introducing a crack in the test specimen and studying its propagation. In case of metals, the testing procedure for fracture toughness is described in IS 10180 and IS 17151. The test procedure for fracture toughness can be simplified as follows:

- Take a sample and prepare a notch in the sample.
- Introduce a crack in the sample at the location of the notch. The crack should be of specified length. In order to control the length of crack, fatigue loading is applied for initial cracks.
- Perform strength test on the cracked sample, in the form of or tension test or bend test.
- Record the deformation, notch opening and load resistance during the test. Notch opening can be recorded using a clip gauge.
- The maximum load is observed and fracture toughness (K_{IC}) is calculated.

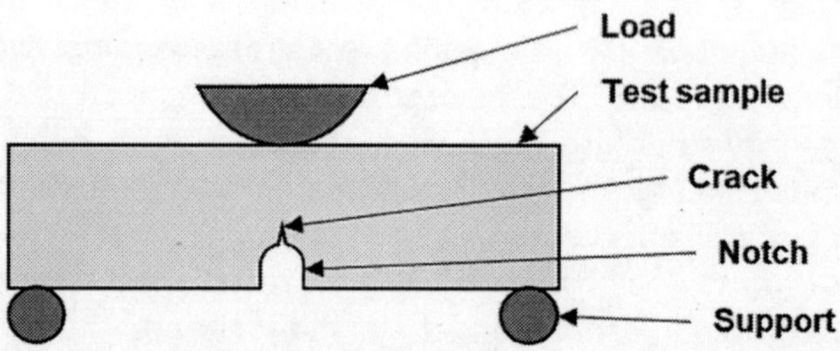

Fig. 4.12: Determination of fracture toughness using bend/flexural test

A material having higher fracture toughness is better at preventing the sudden failure and is considered better in terms of fracture mechanics. Table 4.4 provides the fracture toughness of a few selected materials. It can be seen from Table 4.4 that materials like glass and concrete have very low toughness. These materials will not be able to resist crack propagation, and hence exhibit a sudden failure. On the other hand, stainless steel has very high toughness and can resist crack propagation. It should be kept in mind that low fracture toughness indicates sudden failure, and not brittle failure. Take the case of cast iron and ABS polymer. Cast iron is brittle but has higher toughness than ABS polymer which is ductile.

Table 4.4: Fracture toughness of some selected materials
(Reference: N. Subramanian, Building Materials Testing and Sustainability, Oxford University Press, 2019)

Material	Toughness K_{IC} (MPa.$\sqrt{\text{m}}$)	Material	Toughness K_{IC} (MPa.$\sqrt{\text{m}}$)
Cast iron	33	ABS polymer	3
Al-Ti alloy Al 2024-T351	34	Low-Carbon steel	77
Stainless steel	220	PVC	2.4
Soda-Lime glass	0.76	Concrete	1.19

4.5 FATIGUE TEST

Fatigue tests are performed to measure the performance of materials under dynamic or cyclic loading. Fatigue tests are carried similar to any other strength test, with the exception of loading behaviour. In normal strength test the displacement is gradually increased and load resistance of specimen is measured. On the other hand, in fatigue test loading rate is alternated between two selected stress values for a large number of times, and resulting deformations are measured. The fatigue tests have two key considerations:

1 *Cyclic loads:* Cyclic loads accounts for the upper and lower value of loads between which the loads are being alternated, and the frequency of alternation. In general, the applied loads are less than yield strength, and have same nature of forces, i.e., either compressive or tensile. However, depending up on the application requirement, like stress reversal, the applied loads may be alternated between compression and tension.
2 *Nature of test:* Depending on the application requirements the fatigue test may be performed for different test setups, like compression, tension, flexure and torsion.

Fatigue tests are typically performed for a specified number of load repetitions or till failure. The results are reported in terms of number of cycles required to break the specimen (N) for a maximum applied stress (S). The minimum applied stress is usually adopted as the minimum operational stress, which can be taken as zero if stress reversal is not involved. Typically, the fatigue tests are performed for different maximum stresses to obtain a S-N curve. A sample S-N curve has been shown in Fig 4.13.

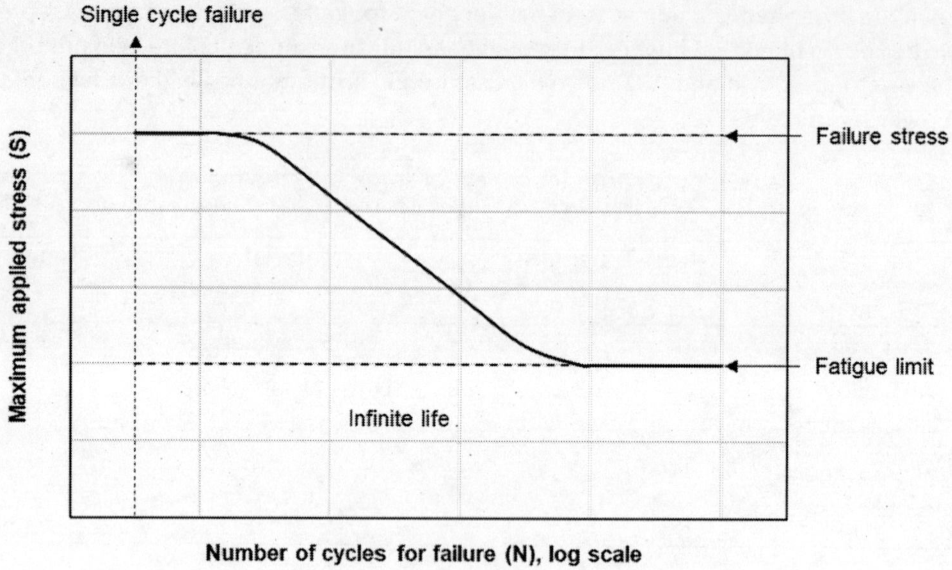

Fig. 4.13: Typical S-N curve for a material

It can be observed from Fig. 4.13 that as the maximum applied stress increases, the material fails in a smaller number of cycles. However, as we decrease the value of maximum applied stress, number of cycles required for failure starts to increase. There comes a point where reducing the maximum applied stress does not cause any failure, and this stress is known as fatigue limit. The point where material reaches the fatigue limit is defined as finite life. Below fatigue limit the material does not fail in fatigue and theoretically shows infinite life.

The fatigue test is typically used to find either of the following:

- Fatigue limit, as maximum permissible stress for an infinite life.
- Maximum permissible stress which can be permitted on the material to support the expected number of cycles.
- Maximum number of cycles a material can support for given applied stress.

4.6 CREEP TEST

Creep test is another variation of strength test where material is subjected to a constant load and change in deformation is observed over a long period of time. It should be noted that deformation under creep occurs at a very slow rate. The creep test is often performed for days, even as long as several years. The creep test essentially requires two specified conditions, i.e., a fixed value of applied loads and a defined temperature range (environmental conditions). Other environmental conditions can also be considered based on the type of material and its intended application. In creep test the loads are applied gradually and the strains are noted throughout the loading and over a long period of time. The results should account for both initial loading strain and creep strain as a function of time. Fig. 4.14 shows a typical representation of creep strain vs. time.

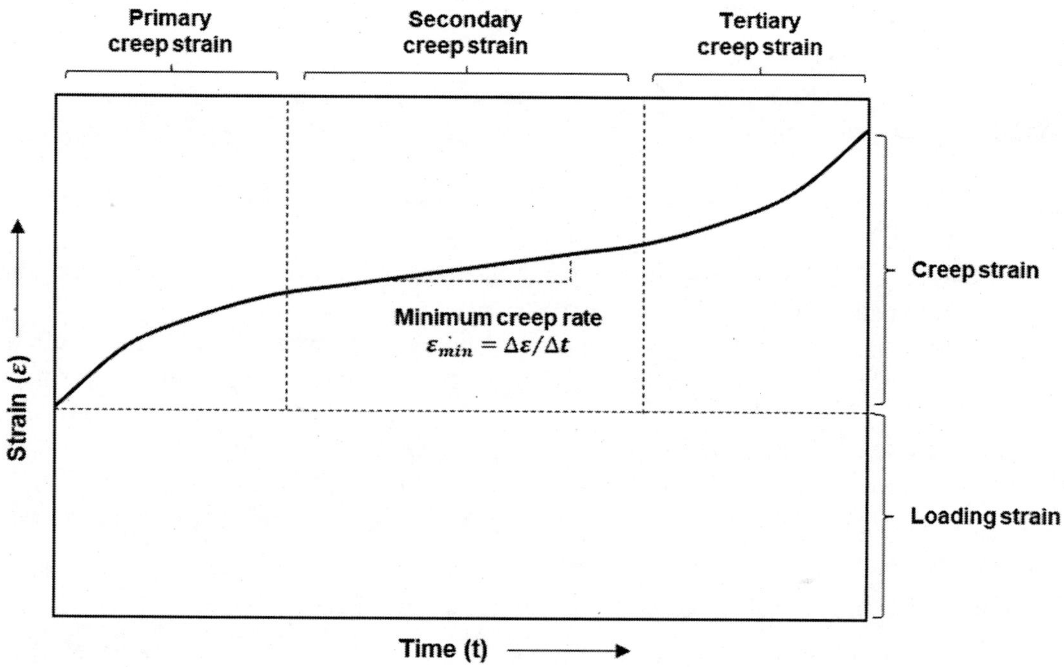

Fig. 4.14: Typical creep strain curve

The creep strain vs. time graph is studied in terms of creep rate, i.e., the rate of change of creep strain with time. During primary phase the creep rate is high but decreases with time. The creep rate reaches a minimum constant value in secondary phase of the creep. An in the third phase the creep again begins to increase with time. It should be noted that the creep rate is a function of applied stress. The test can be repeated over different stress values to observe the various minimum creep rate. Typically, limits are specified on the permissible value of minimum creep strain, which as per IS 3407 is taken as 10^{-7} per hour. The stress at which this value of minimum creep strain rate occurs is known as limiting stress. During field application it is advised that the applied stress should be less than the limiting stress.

4.7 TESTING OF CEMENT

Cement is one of the most commonly used civil engineering material. Unlike most materials like steel, bricks and concrete, the cement is not used directly for civil engineering applications. The cement is used as a binding agent and then contributes to strength application. Therefore, the testing of cement is more important from application and quality control point of view than mechanical characteristics. Different tests are carried out on cement, to check its engineering performance as per the application requirements. In general, the testing of cement can be categorized into two types, i.e., field test(s) and laboratory test(s).

4.7.1 Field tests:

Field tests are used to provide a quick assessment on the quality of cement, in absence of a proper testing facility. Field tests are typically used on site for a qualitative assessment.

(i) Physical property test: Typically, a good quality cement shows a specific set of physical properties, which can be verified as follows,

- On visual inspection, the given sample of the cement should be of uniform grey colour with a light greenish shade.
- When rubbed between fingers, the given sample of the cement should feel smooth to touch.
- On gently crushing the sample, it should break into powders and should be free from air set lumps.
- If the sample of cement is thrown in a bucket of water, it must sink and not float.
- When mixed with water to prepare a thin paste, the cement must feel sticky.
- If the hand is placed inside the bag of cement, it should feel cool on touch, not warm.

(ii) Setting test: In a typical setting test, following steps are used,

- A sample of cement is mixed with water to prepare a thick paste and placed over a glass plate.
- The glass plate, with cement paste, is immersed in water for 24 h.
- After 24 h, the cement paste should be set and does not show any cracks.

(iii) Strength test: In a typical strength test, following steps are used,

- A sample of cement is mixed with water to prepare a thick paste.
- The paste is then used to prepare a block of 25 mm × 25 mm × 200 mm and immersed in water.
- After 7 days, the block is removed from water and placed over two supports that are 150 mm apart (Fig. 4.15).
- A concentrated point load of approximately 340 N (or 34 kg) is applied on the block and checked for failure. A good quality cement will not break or show any cracks.

Setting test and strength test are subjective tests. Results can vary based on various factors like different water content, cement type and degree of compaction. And hence, these tests should only be used as a qualitative assessment; and laboratory tests should be used for determining the actual setting and strength properties.

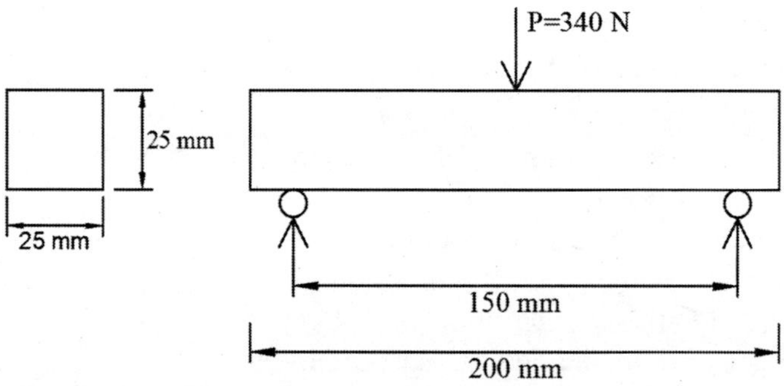

Fig. 4.15: Test set up for strength test of cement block

4.7.2 Laboratory tests

Laboratory tests are used to provide an accurate assessment on the properties of cement, using suitable testing facilities.

(i) Fineness test: Fineness test is carried out to determine the surface area and size of cement particles. The fineness of cement is used to check the degree of grinding. As finer cement particles (small size or large surface area) will have a higher rate of hydration and will increase the rate of setting, strength gain and evaluation of heat. The fineness of the cement can be tested by any of the following methods:

(a) Sieve test: The process of sieve test is described in IS 4031 (part 1). To perform this test, take 10 g of cement sample and place it over sieve having a mesh size of 90 μm. Agitate the sieve by swirling and shaking, which allows the finer particles to pass through the sieve. Break air set lumps (if formed) and continue the sieving till no more particles pass through the sieve. Express the weight of the residue left over the sieve as the percent of cement used in the study, i.e., 10 g. For OPC cement, the residue or coarse sized particles must not be greater than 10%.

(b) Air permeability test: The process of air permeability test is described in IS 4031 (part 2). In this test, first take a known quantity of sample and compact it inside a special cylindrical cell between two filter paper discs. This arrangement is called compacted cement bed. Place the compacted cement bed on the Blaine air permeability apparatus over one end of a manometer. Subsequently, create a low pressure between the compacted cement bed and the manometer, to raise the manometer liquid to a specified height. After this, allow the air to pass through the compacted cement bed, which allows the manometer liquid to fall. The time required for the manometer liquid to drop by a specific height (or pressure change) gives the permeability of compacted cement bed, which is used to determine the surface area of cement particles. For OPC cement, the surface area of cement must not be less than 2250 cm^2/g.

(ii) Specific gravity: The specific gravity of cement is obtained by using Le Chatelier's flask (Fig. 4.16). The specific gravity of Portland cement is typically 3.1. while, specific gravity is not an indication of quality of cement, it plays an important role during the proportioning of concrete mix.

- The process of determining specific gravity is described in IS 4031 (part 11) and can be summarized as follows:
- Fill the Le Chatelier's flask with either water-free kerosene, or naphthalene with a specific gravity not less than 0.7313, to a point on the stem between the 0 ml and 1 ml mark.
- Place the Le Chatelier's flask in a water bath at a constant temperature, typically 20°C, and note the height of the liquid as initial reading.
- At the same temperature as the liquid, gradually add a weighted amount of cement (about 64 g).
- Once all the cement has been added, put the stopper on the flask. Then gently roll the flask in an inclined position or whirl in a horizontal circle to release the air from the cement, until no more air bubbles are visible on the liquid's surface.
- Place the flask back into the water bath, and note the height of the liquid as final reading.
- The volume of liquid displaced by the amount of cement used in the test is indicated by the difference between the initial and final readings.
- The volume of liquid displaced is used to determine the specific gravity of cement using the following equation.

$$Specific\ gravity = \frac{Weight\ of\ cement\ (gm)}{Volume\ of\ liquid\ displaced\ (ml)\ \times\ Density\ of\ water\ (gm/ml)} \quad (4.3)$$

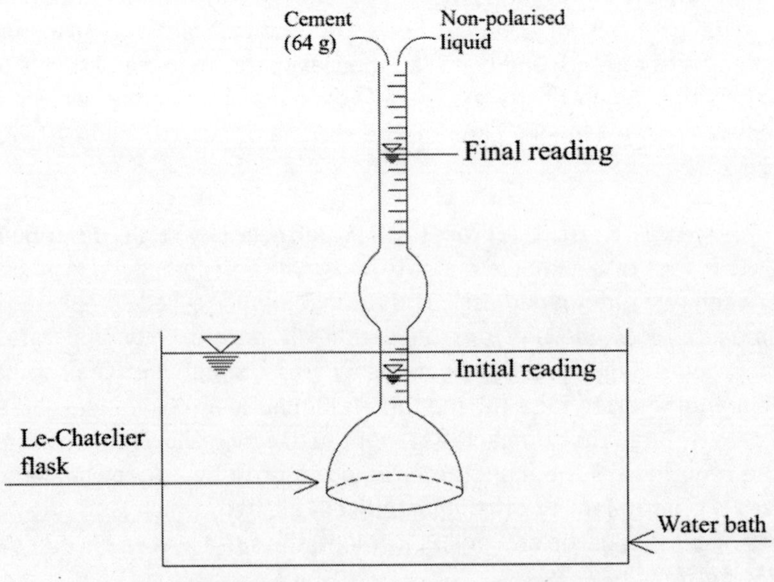

Fig. 4.16: Test set up for specific gravity test

(iii) Consistency Test: To find initial setting time, final setting time, strength and soundness of any given cement sample, a parameter referred to as a standard consistency is required. Consistency test is carried out by Vicat's apparatus (Fig. 4.17). Standard consistency is described as the state of cement paste which allows a plunger with a 10 mm diameter and 50 mm height to enter the mould up to 33-35 mm from the top, or 5-7 mm from the bottom. The purpose of this test is to find the water content which is required to prepare the cement paste of standard consistency. The process of consistency test is described In IS 4031 (part 4). Following steps are performed in this test;

- Weigh a known quantity of cement (typically 500 g) and mix it with a selected amount of water, say 24% by weight, to prepare a cement paste.
- Fill the cement paste in the mould and place it on the Vicat's apparatus.
- Release the plunger over the cement paste and note the depth of penetration.
- Repeat the test with different water content until a penetration of 33-35 mm is observed.
- The water content which gives the penetration depth of 33-35 mm is recognized as the water required for standard consistency and denoted by "*P*".
- Typically, the initial mixing water is kept low and water is increased till the standard consistency is achieved.

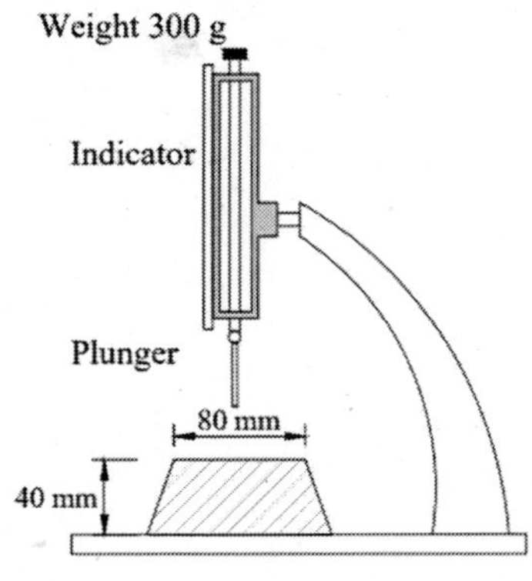

Fig. 4.17: Vicat's apparatus

(iv) Setting times: Setting times are used to define the setting or hardening behaviour of a cement paste. Setting time indicates the amount of time available to perform the initial operations on cement. Setting time of the cement is further classified into following:

(a) Initial setting time: The amount of time that has passed since the cement and water were combined, which causes the paste to start losing its fluidity or workability is known as initial setting time. The test procedure for determining initial setting time is described in IS 4031 (part 5) and is as follows:

- Take about 300 g of cement and mix it with 0.85 P amount of water. Here, P denotes the amount of water required to make a cement paste of standard consistency.

- Attach the square needle to the Vicat plunger and lower it gently to make contact with the surface of the test block and quickly release it.
- Note the depth penetrated by the needle.
- Repeat steps 2 and 3 (preferably every 2 minutes) by moving the needle tip to another location on the surface of the cement paste.
- When the needle penetrates only to a depth of 35±0.5 mm from the top (or 5±0.5 mm from the bottom), the test is complete, and the time from addition of water is noted as initial setting time.
- Initial setting time should not be less than 30 minutes for OPC and 60 minutes for low heat cement.

(b) *Final setting time:* The final setting time is the period that passes between the addition of water to the cement and the point at which the paste has entirely lost its fluidity but has retained enough firmness to withstand a specific amount of pressure. The test procedure is as follows:

- Take about 300 g of cement and make a cement paste of consistency of 0.85 P. The final setting time can be performed in continuation with initial setting time and the same block of cement paste can be used.
- Replace the square needle by a needle with an annular collar.
- Lower the attachment to gently cover the surface of the test block.
- Repeat step 3 at different time intervals (preferably every 15 minutes) by moving the needle tip to another location on the surface of the cement paste.
- The cement is deemed to be fully set if the needle leaves an impression, but the annular collar of the attachment does not. As a result, the paste has grown so tough that a needle cannot pierce it more than 0.5 mm deep.
- The time for the fully set cement from the addition of water is noted as the final setting time. The total time for setting should not exceed 10 hours.

(v) Soundness test: Soundness of cement indicates that the cement paste, once it has set, does not undergo appreciable change in volume and does not cause concrete to crack. The cement having some quantity of free lime, magnesia and excess sulphates undergoes large changes in volume as the time elapses and causes cracks. The soundness of cement indicates the lack or limited quantity of such compounds in cement. The soundness of cement is determined either by 'Le Chatelier's method' or by means of 'Autoclave test', as described in IS 4031 (part 3).

(a) *Le Chatelier's method:* The Le Chatelier's apparatus (Fig. 4.18) consists of a small split cylinder of spring brass. It is 30 mm in diameter and 30 mm high. On either side of the split are attached two indicator arms 165 mm long with pointed ends.

- Take about 100 g of cement and make a cement paste of consistency 0.78 P.
- Fill the cement paste in the split cylinder mould and keep it on a glass plate. Cover the mould on the top with another glass plate.
- Submerge the entire assembly in water at a temperature between 27 ± 2 °C for 24 hours.

- After 24 hours, remove the assembly and measure the separation between the indication points.
- Then, immerse the mould in a water bath. Bring the water bath to boiling point, in 25 to 30 minutes. Keep the mould submerged in boiling water for 3 hours.
- Take the mould out of the water and allow it to cool. Now, measure the separation between the indication points.
- The difference between the two measurements represents the expansion of cement.
- This method detects unsoundness due to free lime only. This testing procedure cannot detect the presence or consequences of an excess of magnesia and calcium sulphate. For OPC, rapid hardening, and low heat Portland cements, the expansion of cement must not be greater than 10 mm by this approach. OPC, Rapid Hardening Cement, Low Heat Cement, PPC can have maximum expansion less than 10 mm whereas high alumina cement and super sulphated cement can have maximum expansion less than 5 mm.

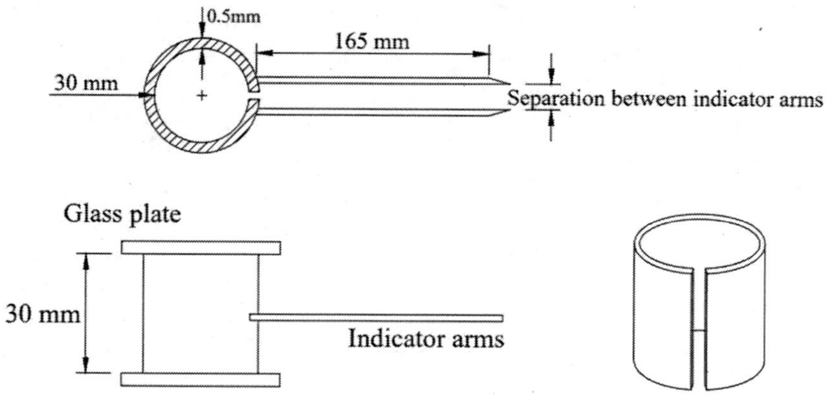

Fig. 4.18: Le-Chatelier's Apparatus

(b) *Autoclave Method:* Typically, a cement with a magnesia content of more than 3% should undergo an autoclave test to determine its soundness. As compared to Le-Chatelier's method, autoclave method can detect both free magnesia and free lime. The autoclave method for soundness of cement can be described as follows:

- Take about 500 g of cement and mix with (P amount of) water to prepare a cement paste of standard consistency.
- Place the cement paste in a mould of size 25 mm × 25 mm × 250 mm.
- Keep the mould in a moist closet or room for 24 hours and then demould the cement paste specimen.
- Measure the length of the specimen, and then place the specimen on a rack inside the autoclave.
- Now, raise the temperature and pressure inside the autoclave at such a rate to bring the gauge pressure of the steam to 21 kg/cm^2 in 1 hour to 1 hour 15 minutes from the time the heat is turned on.

- Hold this pressure for 3 hours.
- Cool the autoclave, and measure the length of the specimen.
- The expansion should not exceed 0.8% for OPC, SRC, PPC, RHC, LHC, Slag cement

(vii) Chemical composition test: Chemical composition of cement is determined using chemical reagents as per IS 4032 or through sophisticated instruments like XRF and XRD. As the test procedure varies significantly for each chemical and is rarely used except by cement manufacturers, individual tests have not been discussed. Based on the results of chemical composition test, the following should be checked for a good quality cement:

- Alumina to iron oxide ratio must have a proportion that is greater than or equal to 0.66.
- Total magnesia must not be greater than 5%
- Total sulphur content must not be greater than 1.5%
- Total loss on ignition must not be greater than 4%
- It is recommended that the ratio of lime to silica, alumina, and iron oxide not exceed 1.02 and not fall below 0.66. This ratio is termed as lime saturation factor

$$0.66 < \frac{Cao - 0.7SO_3}{2.8SiO_2 + 1.2Al_2O_3 + 0.65Fe_2O_3} < 1.02 \quad (4.4)$$

(viii) Compressive strength: The procedure of finding out compressive strength is described in IS 4031 (Part 6). The compressive strength test is similar to any conventional mechanical test and only requires specific guidelines for sample preparation. The test can be performed as follows:

- Take 200 g of cement and 600 g of standard sand (Ennore sand) (i.e., ratio of cement to sand is (1:3) and mix them.
- Add water of quantity (P/4+3.0) % of combined weight of cement and sand, where P is the percentage of water required to produce a cement paste of normal consistency
- This mixing period shouldn't be shorter than 3 minutes and not longer than 4 minutes.
- Immediately, after mixing, fill the mortar in a cube mould of size 7.06 cm. The area of the face of the cube should be 50 cm^2.
- For 24 hours, keep the filled mould at a temperature of 27° ± 2°C and at least 90% relative humidity.
- After 24 hours, remove the mortar cubes from the mould and immerse in fresh, clean water until testing.
- Test three cubes each for compressive strength at the required intervals, typically 1 day, 3 days, 7 days, and 28 days, with the testing period being measured from the packing of mortar in cubes.
- Apply load gradually at a rate of 35 N/mm^2 per minute on cubes by using compressive testing machine. Note the load at which the mortar cubes fail.
- Determine the compressive strength by averaging the strength of the three cubes.
- The compressive strength of a 33 grade OPC at 3 days, 7 days and 28 days should not be less than 16 Mpa, 22 Mpa and 33 Mpa respectively.

(ix) Tensile strength: The tensile strength of cement may be obtained by Briquette test (Fig. 4.19). It should be noted that although tensile strength of cement is not covered in Indian Standards, several of the international testing standards use this test. The Briquette like shape of cement helps in gripping the sample for direct tensile strength test. Typically, the test can be performed as follows:

- Measure cement and sand in a weight-to-weight ratio of 1:3.
- Calculate the percentage of water to be used by the formula (P/5+2.5) % where P is the percentage of water required to produce a paste of standard consistency.
- Fill the mix in the briquette moulds and the surface of the mould is finished with the blade of a trowel.
- Maintain the briquette mould for 24 hours at a temperature of 27 ± 2 °C and in an environment with 90 % humidity.
- Store the briquettes in clean water and remove them prior to testing.
- Compute the average tensile strength of six briquettes.
- Apply the load uniformly and slowly, starting at zero and rising by 0.7 N/mm^2 every 12 seconds.
- After 3 and 7 days, respectively, OPC cement should have a tensile strength of at least 2 Mpa and 2.5 Mpa.

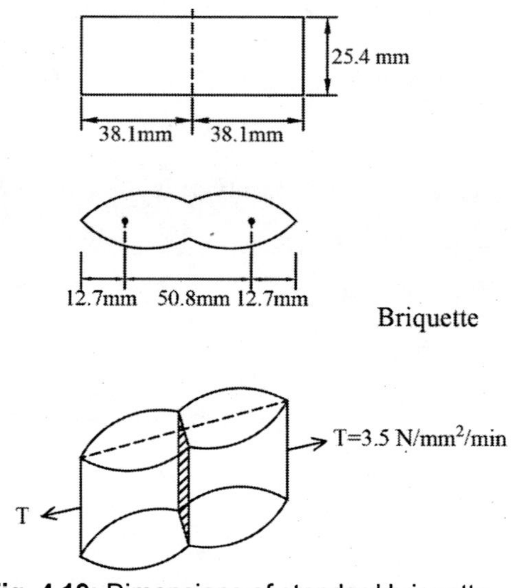

Fig. 4.19: Dimensions of standard briquette

UNIT SUMMARY

Mechanical testing

- Laboratory equipment
- Discussion about working principles
- Preparation of test samples

Different types of tests

- Mechanical tests – strength, impact, fracture toughness, fatigue and creep
- Other important tests on cement
- Testing and evaluation procedures

Interpretation of test results

- Recording of essential data.
- Calculation and identification of key results from the data.
- Interpretation of material characteristics based on the available data.

EXERCISES

Multiple Choice Questions

4.1 Which of the following properties can be tested on a universal testing machine:

(a) Impact energy under tension (b) Resilience under flexure

(c) Creep strain rate under torsion (d) All of the above

4.2 Impact test can be performed using:

(a) Drop impact test setup (b) Izod test setup

(c) Charpy test setup (d) All of the above

4.3 The name FeE200SP55 denotes:

(a) Yield strength of 200 MPa and ultimate strength of 255 MPa

(b) Ultimate strength of 200 MPa and yield strength of 0.55×200 MPa

(c) Yield strength of 200 MPa only and 55 represents degree of purity

(d) Ultimate strength of 200 MPa only and 55 represents degree of purity

4.4 Concrete is prepared using the mix design guidelines listed in

(a) IS 456 (b) IS 10262

(c) IS 516 (d) IS 4031

4.5 In creep test, limiting stress is defined as the stress at which minimum creep strain rate is

(a) 10^{-3} percent per hour (b) 10^{-4} percent per hour

(c) 10^{-5} percent per hour (d) 10^{-6} percent per hour

4.6 Below transition temperature yield strength and fracture toughness of steel

(a) Both increases (b) Both decrease

(c) Yield strength increases but fracture toughness decreases (d) Yield strength decreases but fracture toughness increases

4.7 What is not visible in the typical stress strain curve of brittle material

(a) Plastic deformation (b) Yield point

(c) Strain hardening (d) All of the above

4.8 Which of the following metal and its alloys use a numeric naming system

(a) Aluminium (b) Manganese

(c) Copper (d) Steel

4.9 Which of the following are true:

(i) A material with higher fracture toughness has higher ductility, and vice versa

(ii) Minimum creep rate is observed during the primary phase of creep strain

(a) Both are true (b) Both are false

(c) Only (i) is true (d) Only (ii) is true

4.10 To check the possibility of expansion crack in cement concrete, following test should be performed

(a) Soundness test (b) Creep test

(c) Consistency test (d) Fracture toughness test

Answers of Multiple Choice Questions

Answers of Multiple Choice Questions
4.1 (b), 4.2 (d), 4.3 (c), 4.4 (b), 4.5 (c), 4.6 (c), 4.7 (d), 4.8 (a), 4.9 (b), 4.10 (a)

Short and Long Answer Type Questions

4.1 Write a short note on preparation of sample for tension test of steel.

4.2 What are the steps involved in setting time of cement. Describe in details.

4.3 Describe and differentiate the method of sample preparation for mechanical testing of cement, mortar, concrete and asphalt.

4.4 What is fatigue limit and describe its relevance to infinite material life.

4.5 How can we determine the transition temperature of steel? Describe.

4.6 What is embrittlement.

4.7 Write a short note on field tests of cement.

Numerical Problems

4.1 In a compression strength test, using a concrete of size 15 × 15 × 15 cm, the load vs. deflection graph showed proportional limit up to 30 kN and 2 mm. Calculate the initial modulus of elasticity of concrete. Assuming the material yielded at this point only then calculate the resilience of the given material.

4.2 During a pendulum type impact test, a hammer of weight 100 kg is swung using a rod of 1m. The hammer was dropped from an angle of 90° from the vertical, strikes the sample at 0° from the vertical and reaches an angle of 60° from the vertical. Calculate the impact energy absorbed by the test specimen.

4.3 Calculate the proportion of raw ingredients required for determining the compressive strength of cement. The standard consistency may be adopted as 27%.

PRACTICAL

Relevant practical for this unit have been included within the unit itself.

KNOW MORE

Durability

Durability is an important characteristic of any engineering material. Durability is defined as the ability of a material to perform satisfactorily under different exposure conditions over long periods of time. For example, steel corrodes in presence of moisture and loses its mechanical characteristics, which may cause the structure to fail. A durable material is able to retain its desired mechanical characteristics for the expected service life. Therefore, it is important to understand the durability of engineering materials under various operational requirements.

Durability is defined in terms of loss in other characteristics and as such durability has no independent testing procedure. Durability is typically determined by recreating the exposure conditions in a controlled environment and repeating the tests for mechanical characteristics at different stages. For example, in case corrosion steel is corroded electrochemically and at specified corrosion level tested for mechanical strength.

Therefore, one should keep in mind that while durability is not discussed independently, it is an important characteristic of any engineering material. One should have an understanding of the various exposure conditions and material is expected to encounter during its application. Furthermore, performing standard tests after simulating the exposure condition indicates the durability of the materials. A durable engineering material is less likely to fail during the service life and should be adopted in construction.

REFERENCES AND SUGGESTED READINGS

1. K. Komvopoulos, *Mechanical Testing of Engineering Materials*, Cognella, 2017
2. M. R. Islam, *Civil Engineering materials – Introduction and laboratory testing,* CRC Press, 2020
3. J. Roesler, H. Harders, M. Baeker, *Mechanical Behaviour of Engineering Materials*, Springer, 2007
4. N. E. Dowling, S. L. Kampe, M. Kral, *Mechanical Behaviour of Materials*, Pearson, 2019
5. V. Raghavan, *Materials Science and Engineering*, PHI Learning Pvt. Ltd., 2021
6. N. Subramanian, *Building Materials Testing and Sustainability*, Oxford University Press, 2019.
7. J. P. Patel and G. H. Upadhyay, *Material Science*, Atul Prakashan, 2001.

Dynamic QR Code for Further Reading

Intentionally left blank!

CO AND PO ATTAINMENT TABLE

Course outcomes (COs) for this course can be mapped with the programme outcomes (POs) after the completion of the course and a correlation can be made for the attainment of POs to analyze the gap. After proper analysis of the gap in the attainment of POs necessary measures can be taken to overcome the gaps.

Table for CO and PO attainment

Course Outcomes	Attainment of Programme Outcomes *(1- Weak Correlation; 2- Medium correlation; 3- Strong Correlation)*											
	PO-1	PO-2	PO-3	PO-4	PO-5	PO-6	PO-7	PO-8	PO-9	PO-10	PO-11	PO-12
CO-1												
CO-2												
CO-3												
CO-4												
CO-5												
CO-6												

The data filled in the above table can be used for gap analysis.

INDEX